Learning Materials in Biosciences

W0230353

Learning Materials in Biosciences textbooks compactly and concisely discuss a specific biological, biomedical, biochemical, bioengineering or cell biologic topic. The textbooks in this series are based on lectures for upper-level undergraduates, master's and graduate students, presented and written by authoritative figures in the field at leading universities around the globe.

The titles are organized to guide the reader to a deeper understanding of the concepts covered.

Each textbook provides readers with fundamental insights into the subject and prepares them to independently pursue further thinking and research on the topic. Colored figures, step-by-step protocols and take-home messages offer an accessible approach to learning and understanding.

In addition to being designed to benefit students, Learning Materials textbooks represent a valuable tool for lecturers and teachers, helping them to prepare their own respective coursework.

More information about this series at http://www.springernature.com/series/15430

Eva-Kathrin Ehmoser-Sinner
Cherng-Wen Darren Tan

Lessons on Synthetic Bioarchitectures

Interaction of Living Matter with Synthetic Structural Analogues

 Springer

Eva-Kathrin Ehmoser-Sinner
Institute for Synthetic Bioarchitectures
University of Natural Resources
and Life Sciences
Vienna, Austria

Cherng-Wen Darren Tan
Institute for Synthetic Bioarchitectures
University of Natural Resources
and Life Sciences
Vienna, Austria

ISSN 2509-6125   ISSN 2509-6133   (electronic)
Learning Materials in Biosciences
ISBN 978-3-319-73122-3   ISBN 978-3-319-73123-0   (eBook)
https://doi.org/10.1007/978-3-319-73123-0

Library of Congress Control Number: 2017964085

© Springer International Publishing AG 2018
This work is subject to copyright. All rights are reserved by the Publisher, whether the whole or part of the material is concerned, specifically the rights of translation, reprinting, reuse of illustrations, recitation, broadcasting, reproduction on microfilms or in any other physical way, and transmission or information storage and retrieval, electronic adaptation, computer software, or by similar or dissimilar methodology now known or hereafter developed.
The use of general descriptive names, registered names, trademarks, service marks, etc. in this publication does not imply, even in the absence of a specific statement, that such names are exempt from the relevant protective laws and regulations and therefore free for general use.
The publisher, the authors and the editors are safe to assume that the advice and information in this book are believed to be true and accurate at the date of publication. Neither the publisher nor the authors or the editors give a warranty, express or implied, with respect to the material contained herein or for any errors or omissions that may have been made. The publisher remains neutral with regard to jurisdictional claims in published maps and institutional affiliations.

Printed on acid-free paper

This Springer imprint is published by Springer Nature
The registered company is Springer International Publishing AG
The registered company address is: Gewerbestrasse 11, 6330 Cham, Switzerland

Contents

Introduction

Electronic supplementary material The online version of this article
(https://doi.org/10.1007/978-3-319-73123-0_1) contains supplementary material,
which is available to authorized users.

© Springer International Publishing AG 2018
E.-K. Ehmoser-Sinner, C.-W. D. Tan, *Lessons on Synthetic Bioarchitectures*, Learning Materials
in Biosciences, https://doi.org/10.1007/978-3-319-73123-0_1

1

1.1 Long-Term Vision and Objectives

The objective of synthetic bioarchitectures as a field of research cannot be confined yet as it belongs to the converging sciences, still emerging; however, let us foresee one of the most relevant objectives of this field: the communication of life with synthetic matter.

What can we learn by talking to nature in the language of molecules? We can interfere with biological pathways in a much more "compatible" format than has ever been possible before.

For example, thinking about chemotherapy we might apply the German saying: "den Teufel mit dem Beelzebub austreiben"—which means that chemotherapy is about trading off: lacking specific tumor markers results in the attempt to stop proliferation in general and the result appears as treating "bad with similar bad": we kill various cells in the course of chemotherapy and eventually we succeed by hitting cancerous cells harder than benign tissue. The side effects are of course enormous and undesired.

But imagine a novel way to address such cancerous tissue. What a difference it would make if by means of synthetic biology—namely, bottom-up approaches—we were able to synthesize "communicators," talking only to the desired cells without toxifying them—rather, "convincing" them to get back into the healthy regulated routines of benign tissue. It still sounds naïve; however, we have come a long way in "understanding" biological architectures (◘ Fig. 1.1).

Can we think of a cell as communicateable object, where synthetic assemblies might integrate meaningfully into regualtory processes of diseased pathways and by this, cure, dysfunction?

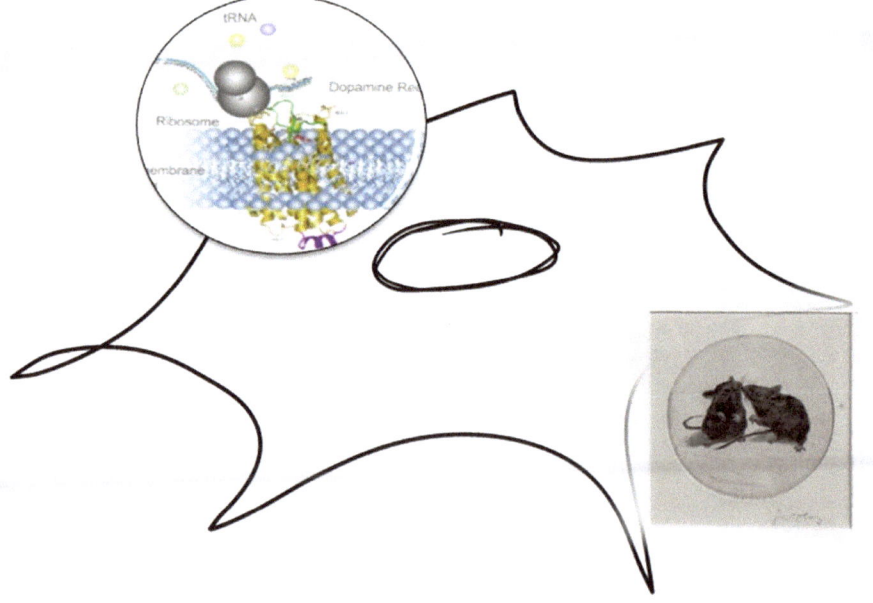

◘ **Fig. 1.1** Cartoon of a (eukaryotic) cell with a graphic inset of an artificial bioarchitecture—namely a membrane protein, which might interfere with a diseased cell. This desired artificial assembly is available in vitro, made by the ribosomal complex with all the compounds needed (translocon machinery, chaperones, etc.) and energetic boundary conditions involved; however, instead of targeting the endoplasmic reticulum, as an example, colloidal membrane architecture can be addressed (e.g., membrane disks). On the other side of the cartoon, the whole organism is depicted, represented by two mice, eventually being cured by the introduction of such "synthetic assemblies"

Future vision: 'Communication'
between materials and living cells on all levels

◘ Fig. 1.2 As time will tell, eventually, synthetic bioarchitectures will be targeted by scientists as novel therapeutics. (Composition images courtesy of D. Miklavcic, Ljubljana University, Slovenia; Tarek Mounir, CNRS, France; Ute Reuning, Technische Universität München, Germany)

We like to present, as an example, clinically relevant membrane proteins as such communicators, which are integrated into biocompatible polymeric islands as shuttle systems. Such orthogonal therapies would allow restoration of function at endogenous receptor levels. However, biofunctionalized surfaces, biohybrids, and novel gene-coding strategies point in the very same direction: bioarchitectures as novel approaches in communication with nature. Such fundamentally novel materials will be highly attractive for the pharmaceutical industry and molecular medicine. We present such pioneering technologies, which provide great opportunities for developing next-generation protein therapies in order to address fatal membrane protein–related diseases (◘ Figs. 1.2 and 1.3).

■ **Objectives for Students**
1. Relevance of the topic
2. Overview of the content
3. Connections with established disciplines
4. Visions and perspectives

1

Biomaterials : The Building Blocks of Life

Structure

Function

Material Properties

- we try to 'understand' the building blocks of life
- we try to 'mimick' and 'arrange' the building blocks of life into integrated, functional architectures

■ **Fig. 1.3** Depicting the connection between structure and function and their consequence: material properties. And in the center: insight into the beautiful "functional structure of the nautilus shell". This Wikipedia and Wikimedia Commons image is from the user Chris 73 and is freely available at https://commons.wikimedia.org/wiki/File:NautilusCutawayLogarithmicSpiral.jpg under the creative commons cc-by-sa 3.0 license

■ **Expected Outcomes**
1. Students will know the term "synthetic bioarchitectures" with respect to bottom-up–top-down approaches in communicating with nature via materials/surfaces.
2. Relation to the "Roter Faden" of the book.
3. The history and the intra- and interdisciplinary future of such converging technology.
4. Putative goals and achievements in the field: the magic riddle of "regulation" in the context of cancer research.

1.2 Synthetic Biology and Synthetic Bioarchitectures

"Synthetic biology" is a term describing the attempt to synthesize, manipulate, and—first of all—understand nature.

We know about proteins, lipids, and carbohydrates as the building blocks of life. As functional architectures, the building blocks of life have revealed impressive properties in the course of research that has even elucidated electron spin responses in biomaterials. In synthetic biology, several attempts have been made to use nature as a blueprint—for example, for bionic approaches on an engineering level—however, the closer

First time, the term "Synthesis" and "Biology" was put together: by Stéphane Leduc

'Il existe, pour expliquer les phénomènes de la nature, deux méthodes : le Mysticisme et le Physicisme. Le physicisme est la méthode des sciences physiques, le mysticisme règne encore sur la biologie.'

ETUDES DE BIOPHYSIQUE

LA BIOLOGIE SYNTHÉTIQUE

PAR

STEPHANE LEDUC

AVEC 111 FIGURES DANS LE TEXTE

A. POINAT, ÉDITEUR

BOULEVARD SAINT MICHEL 4 PARIS

1912

☐ **Fig. 1.4** If we take a biomaterial in the focus of synthetic bioarchitectures, we have to view the composition of living objects from the perspectives of chemistry, physics, and biology; only then might we reveal the structural–functional relationships and, as a consequence, we might be able to control and eventually mimic the material properties presented by nature

methodological approaches have come to the level of atomic resolution of functionalized assemblies in living cells, the more subtle and ambitious scientists have become worldwide about the vision of mimicking nature (☐ Fig. 1.4).

Going back in history, the term "La Biologie Synthetique" was coined by a French scientist named Stephane Leduc (see ☐ Fig. 1.5). It was a commitment to phenomenologically driven aspects of biology, which in contrast to physics were annotated as being of a "mystical" nature—too complex to ever be understood and described by laws and numbers.

In this context, the origin of attempts to understand and manipulate biology was in alchemy—a descriptive view of our world—attempting to transform the elements of the periodic system and by this to "control" nature.

If we look back in history, the extent to which scientists at the respective times with respective methods were able to understand interaction and relevant processes even on the atomistic level is impressive. For the topic of synthetic bioarchitectures, it became interesting after Boyle, Dalton, and Newton set out the atomistic model, when the periodic system was anticipated by Berzelius, and the elements as components of our world were able to be put on one sheet of paper; even though it was clear that some were still

1

Boyle - (1627–1692)   Newton - (1642–1726)   Dalton - (1766–1844)

And then....

…as a history moves on, alchemy turned into chemistry
and in human followed the idea of atoms being turned into
molecules (lat. moles, f.: mass, molecula: small mass)

Jöns Jakob Berzelius, 1779–1848

🔲 **Fig. 1.5** The key researchers in transforming Alchemy into modern chemistry: the ground work for the periodic system

When Chemistry and Biochemistry meets...

Friedrich Wöhler, 1800–1880

🔲 **Fig. 1.6** Friedrich Wöhler: a capable student crossing borders from synthetic and 'biogenic' materials

missing, the consecutive order of elements according to their mass, defining their properties and interactive capacity, became clear (see 🔲 Fig. 1.6).

However, Berzelius's own student, Wöhler, already thought beyond the border of the periodic system as he performed a transformation of ammonium cyanate into urea by mere heating. This observation was interpreted by him as a transformation from an "unliving" material into a very much "living" material—insofar as he claimed rightfully to have crossed the borders between the inorganic and organic worlds and that there would be

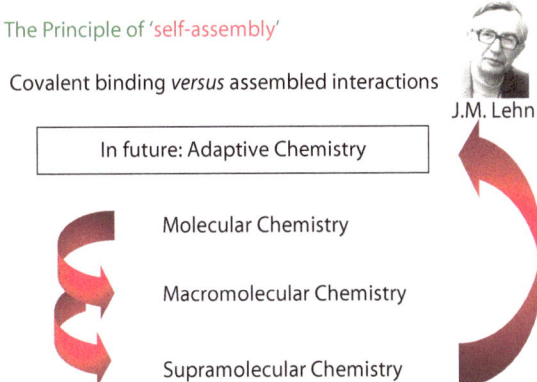

Fig. 1.7 Interpreted from Jean Marie Lehn as he proposes in his presentation at the Journal Angewandte Chemie at its anniversary in Berlin (2012): today's chemistry might become 'adaptive'

connections between elements, which were not describable by the mere periodic system. The letter he wrote to Berzelius, his teacher, is worth reading to see his beautiful spirit of respect and joy about his discovery. And maybe even more acknowledgeable is the response of Berzelius as a supportive and sovereign mentor, even though his student—to a certain extent—had diminished the relevance of his own life's work: the periodic system of the elements as a "sorting table" (see ▶ http://www.chemieunterricht.de/dc2/tip/brief.htm for the original text of the letter and our attached translation into the English language).

Without transition possibilities among the elements themselves, however, as Wöhler demonstrated on the molecular level, materials can be transformed into different combinations and thus the borders between the living and nonliving worlds are not as narrow as was claimed before his observation.

At the beginning of the eighteenth century, it was high time for great discoveries in chemistry, especially in regard to the era of industrialization. The finding of polymeric reactions indeed paved the way for generating materials, which were quite tedious to isolate from natural sources or even impossible to find in nature.

When we think about aniline, the finding of Wöhler took place in the context of an era of identification of new materials with high commercial potential: polymer chemistry. The famous Staudinger reaction took place by condensation of small building blocks—monomers—into polymeric materials, thus the first "plastics" were achieved (■ Fig. 1.7).

The idea of Wöhler—namely, transforming nature's materials easily into controllable, cost-effective, daily life products—was an intriguing thought. Indigo, formerly isolated from plants, provided a perfect example of the potential of synthetic biobased chemistry. In conventional synthetic chemistry, macromolecules are formed by transforming starting molecules into the desired macromolecular material in laboratory conditions by pushing against equilibrium constants and often orchestrating solvent conditions in order to achieve the desired product. As a chemical layman, successful synthesis strategies look to me like cartoons of a spider network, as bypassing the obstacles of accessibility, reaction conditions and, in the end, stability and purity are the factors of turning nonworking into working chemistry. Synthesis blueprints from nature are already guiding open-minded chemists into new spheres of combinatorial organic chemistry, often as a consequence of intensive collaboration with medicine. Often, questions from molecular medicine are answered with strategies and ideas from peptide biochemistry; as an ongoing example.

1

Indigo an example of Bio-inspired synthesis.....

- **Indigo dye** is the dye of blue jeans, for example. It is a perfect example for a paradigm in man made matter, which – of course – is made after mother natures blueprint.

past

future

We might go back to natural sources in some cases!

Production of Indigo dye in a
BASF plant (1890)

- Aniline is THE precursor for synthetic indigo – AND for polyurethan....

▣ Fig. 1.8 The dye "indigo" originally was isolated from the plant named indigofera. However, the synthetic version of indigo was developed as the world needed much more indigo and chemical synthesis based on Aniline precursors enabled a less expensive source. However, the 'blueprint' of this powerful dye came from nature

In the future, we assume that there might not be such a difference between synthetic chemistry and biobased chemistry, as there are already some channels and valves in between, where complex starting blocks are already consequences of the synthesis performed by "active little house elves" in the lab.

No, it is not you, dear students, to which we refer, but bacterial species, which build reliably and reproducibly peptidic/lipid or carbohydrate composites, which can be isolated and further purified in conventional biotechnological down-processing for synthesis reactions.

The advance from macromolecules into supramolecular chemistry is already considered a breakthrough in history—these days the adventure goes on as J.M. Lehn developed the concept of self-assembly (see ▣ Fig. 1.8), and the borders between "man-made" and nature-derived materials seem to fade out once more. As adaptive materials respond to their environment on the molecular scale, what difference is there between an enzyme finding its substrate and a macromolecule with catalytic properties being tunable via pH changes?

The underlying processes in molecular recognition are the same in both cases, and no difference can be found in the consequences of molecular transformations.

Still it is relevant to understand the origin of materials derived from nature, even as we buy them from catalogs in the lab; we should know well the difference, for example, between synthetic peptidic materials and isolated ones, as impurities and stability are major concerns with bioderived materials!

It is not only about impurities being present; it is also about stability and chemical robustness. Materials from biological sources need to be handled accordingly—they are often temperature sensitive and sensible in their structural–functional integrity. This is often ignored when such materials are employed in the interface between disciplines—antibody materials are a prominent example. Antibodies are famous as precise recognition labels and as such are often part of imaging and tracing experiments on the molecular scale. However, they are quite complex biological entities with a limited time of activity and inherent (though often ignored) impurity content as they are isolated from a cell culture supernatant.

The example of antibodies as "major players" in molecular recognition shows the limitation and strength of natural-derived materials.

This example is very useful to show the parallel development of synthetic bioarchitectures as a subfield of synthetic biology; in certain aspects, history repeats itself. As chemistry started from defining the compounds and elements, moving forward to macromolecular and supramolecular chemistry, the field of biomaterials will follow such developments—with the example of antibodies, the field started with defining the material of such protein molecules, the structural features, and modification possibilities. At present, researchers think about artificially copying the binding characteristics of antibodies, either by chemical modification of the present antibodies or by employing alternative materials, such as nucleic acids, aiming to preserve the outstanding capacity in molecular recognition of monoclonal antibodies and at the same time replace the inherent drawbacks with regard to stability, reproducibility, and animal origin.

In this regard, synthetic bioarchitectures seem to follow the field of chemistry with an approximate 30-year gap. However, as already established products, especially based on fossil fuels, are in everybody's hands and minds, it seems difficult to repeat the momentum of the "chemical revolution" from 100 years ago.

Introducing cost- and resource-effective alternatives seems more likely to describe the actual and ongoing movement of synthetic biology in everyday products. Funding schemes, opting for engineered, rational design of commercial alternatives to fossil fuel-based products, are already active and will contribute to hunting down "easy to establish" biobased products, which hopefully will make their name to finally become interesting "stakeholder's darlings."

Why is this a favorable scenario? In our world, the driving momentum will only happen when the common incentives of money and/or power are in place. In research, one can look out for "biomimicking concepts"; impulses from self-organization concepts—known for a long time—are at present being "reinvented" to justify projects in various fields. In synthetic bioarchitectures, we are obliged to start from Feynman's statement, "What I cannot create, I do not understand"—e.g., in his spirit of being able to solve every problem that has been solved and continue with systematic steps toward sustainable, integrated, and intelligible (responsive) functional biomaterials. In summary, this sounds to us like synthetic bioarchitectures.

The putative goals and achievements in this field depend very much on the exact example of choice. In today's biotechnology, the investment in large-sized fermenters is an interesting example to show just one potential of synthetic bioarchitectures: materials from natural sources provide the backbone source for any material processed from our industry. In the beginning of each process, there is a contribution from nature, which in most cases is not even known to the consumer. A "breakthrough invention" from the field of microbiology was the development of single organism–derived cultivation (fermenta-

1

tion), enabling control and foresight in (productive) metabolism. The principle behind it is quite old: knowing the optimal conditions to grow a specific (desired) organism was the ultimate start for human-controlled production of cheese, beer, wine, and many more goodies, where we need biochemical input from the "little helpers" around us—namely bacteria, yeast, and fungi.

In large-scale fermentation, one can find the organism of interest in a monoculture environment, optimized for the production of commercial compounds. In some cases, this concept has taken care of the synthetic chemistry procedures—at least partially—by providing high-purity and cost-effective building blocks of various (industrial) products. Such "organism-based" (or let us call it "cell-based") synthesis still bears some risks and drawbacks, as living species naturally change—for example, as mutations continuously occur. Knowledge about desired biochemical pathways, the enzymes involved, engineering of proteins, and the regulatory interactions between and within microorganisms on very complex levels has been elucidated in the research field of systems biology in the past decades. Already some examples have been applied to address real-world problems, such as synthesis from interesting compounds, engineering of therapeutics (e.g., humanization of antibodies), and green algae turning wastewater into biofuel, just to name a few "dreams" that are about to become reality. One can think about systems biology as a field in which the blueprints of life are elucidated—this is naturally a very important inspiration for any "synthetic" architecture, deriving or making use of such natural building blocks.

In summary, it will be knowledge about molecular interactions and structures of living species that will transform the relevance of synthetic bioarchitectures as an inherent consequence of the idea of synthetic biology into research questions directed toward sustainable, economical, logical, and responsible biotechnological research and industry, which will optimistically be quite integrated into our ethical and societal networks.

Further Reading

Beales PA, Khan S, Muench SP, Jeuken LJC (2017) Durable vesicles for reconstitution of membrane proteins in biotechnology. Biochem Soc Trans 45(1):15–26

Glass JI, Hutchison CA III, Smith HO, Venter JC (2009) A systems biology *tour de force* for a near-minimal bacterium. Mol Syst Biol 5:330. https://doi.org/10.1038/msb.2009.89

Kuruma Y, Stano P, Ueda T, Luisi PL (2009) A synthetic biology approach to the construction of membrane proteins in semi-synthetic minimal cells. Biochim Biophys Acta 1788(2):567–574

Langton MJ, Keymeulen F, Ciaccia M, Williams NH, Hunter CA (2017) Controlled membrane translocation provides a mechanism for signal transduction and amplification. Nat Chem 9(5):426–430

Luisi PL (2007) The emergence of life: from chemical origins to synthetic biology, 1st edn. Cambridge University Press, Cambridge

Malinova V, Nallani M, Meier WP, Sinner EK (2012) Synthetic biology, inspired by synthetic chemistry. FEBS Lett 586(15):2146–2156

Noireaux V, Bar-Ziv R, Godefroy J, Salman H, Libchaber A (2005) Toward an artificial cell based on gene expression in vesicles. Phys Biol 2(3):P1–P8

The Minimal Cell

© Springer International Publishing AG 2018
E.-K. Ehmoser-Sinner, C.-W. D. Tan, *Lessons on Synthetic Bioarchitectures*, Learning Materials
in Biosciences, https://doi.org/10.1007/978-3-319-73123-0_2

2

What You Will Learn in This Chapter

In this chapter, we will present the concept of the minimal cell and examine the reasons for this being a research goal. We will then look at two different approaches that researchers have adopted in order to create the minimal cell. We will also highlight the limits of each method and emphasize the need for the minimal cell to be clearly defined. We will define the minimal cell as any minimal system of interacting molecules that is capable of showing signs of life. By presenting several definitions of life, we will show that conventional definitions do not give practical objectives for the creation of a minimal cell. Autopoiesis is offered as an alternative, functional definition.

2.1 The Minimal Cell

There is a prevailing idea that the complexity of cells we see is the product of evolutionary processes, much like speciation. Underlying this idea is the suggestion that all cells are modifications of what is called a minimal cell. The minimal cell ought to be the simplest collection of interacting molecules that can show signs of cellular life, under specific environmental conditions. Simply put, it is the simplest possible form of cellular life, under those conditions.

While it is uncertain whether the minimal cell ever existed, or is still in existence, in the natural environment, this idea has spurred researchers into seeking to identify, or even create, the minimal cell. They do this for two main reasons. First of all, studying what constitutes a minimal cell would provide insight into the deeper principles of cellular evolution. That is, it would allow us to understand what more it will take to change simple cells into more complex ones. Secondly, a minimal cell can be augmented in various ways to confer new functions upon it. If successful, this will allow researchers to create a new and wider set of cellular tools to solve biological, medical, and environmental problems. This is like reducing a car to just the frame, axles, engine, and four wheels. On this simple base, one could build different chassis to produce cars of different types—an ambulance, a racing car, or a sports utility vehicle, say. You might even build a chassis to produce a car that can do what cars have never done before, perhaps to fly.

There are basically two main strategies employed in the search for the minimal cell: the top-down approach and the bottom-up approach. The top-down approach begins with an existing complex cell (◘ Fig. 2.1). We then try to remove components of the cell and observe the impact of this act. If the cell stays alive, then the part removed is probably not essential to life, under those environmental conditions.

We keep doing this until we have the minimal set of components, fewer than with which, life is no longer possible. This minimal set would be the minimal cell. We will discuss the work of the J. Craig Venter Institute as an example of this approach. The bottom-up approach addresses the challenge from the opposite direction (◘ Fig. 2.2).

Here, basic materials are put together rationally in attempts to reconstruct or mimic biological structures and behavior that might lead to cellular life. This is the sort of approach adopted by those working on chemical autopoiesis, as we will see later.

In both cases, the end point is the minimal cell. However, unless this end point is clearly defined, it will be difficult to know when the minimal cell has been produced. The major question that allows us to define the minimal cell is, "What is life?"

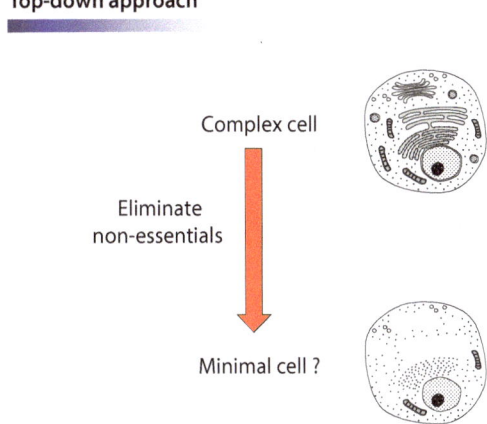

Fig. 2.1 An example of the top-down approach. A cell is systematically stripped of its components in order to attain a desired structure

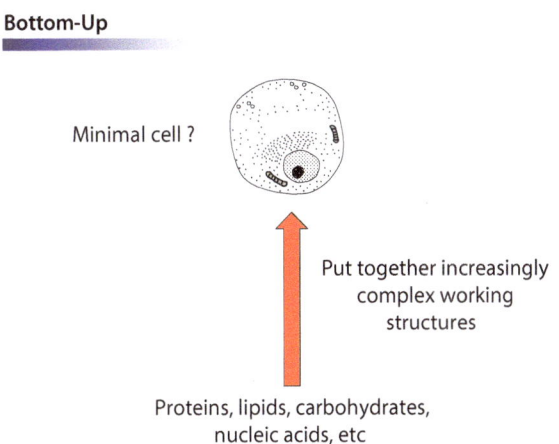

Fig. 2.2 An example of the bottom-up approach. Biomolecules can be brought together to form complex structures such as membranes and simple organelles. These, in turn, can be used to construct even more complex systems, such as a cell

2.2 Defining the Minimal Cell

This is a deeply philosophical question, like so many others that have plagued mankind through the ages. What makes this question so hard to answer is that we do not have a consensual definition for what life is. It is defined in different ways by different people with different backgrounds and different value systems. Some people adopt the same attitude that vitalism does toward the molecules of life—that life is somehow special, is outside the understanding of science, and can never be created artificially.

For the purpose of creating the minimal cell, what we need is a *functional* definition of life—one that is simple, and can be understood in terms of the basic principles of physics and chemistry. If we have such a definition, we can study and understand life more precisely. More than that, it would make the creation of artificial life possible.

When presented with a random collection of items, one can readily agree that the fly, the tree, and the mushroom are living, while the radio, the computer, and the moon are

2

nonliving. Even the single-celled amoeba is considered alive. But things get hazy when we look at viruses. Are viruses alive? There are some who consider them so, and they have their reasons. Others consider their dependence on a host to be proof they are not living. However, if dependence on a host is a contraindication, then should we not also consider the plasmodial parasite—or even a human embryo—nonliving? We need a more pertinent definition that will resolve such conundrums.

This need has been addressed by many philosopher-scientists. The thoughts of Erwin Schrödinger—as outlined in his book *What is Life?*—have had a great impact on both the prevailing definition of life and how people perceive the nature of life. To begin with, he stressed that life consists of phenomena that have to adhere to the laws of nature. This way, they can be understood using science and chemistry. This essentially demystifies life and makes it understandable. Schrödinger's other contribution was to propose that it was an aperiodic crystalline molecule that encoded life—that is, that such a crystal contained all the information necessary for the construction of living systems. If this were true, and if one could find and study such a molecule, one might learn what basic components are necessary for life in general. In doing so, he challenged the prevailing expectation that genetic material was proteinaceous. The year was 1944, close to the end of the Second World War.

This idea was vindicated with the publication of Watson and Crick's seminal *Nature* paper describing the structure of DNA in 1953. The paper itself focuses mainly on the structure and chemistry of nucleic acids. What was significant in Watson and Crick's paper was their description of base pairing and its crucial role in defining DNA structure. However, the truly visionary element of this paper was a comment made near the end: "It has not escaped our notice that the specific pairing we have postulated immediately suggests a possible copying mechanism for the genetic material." With this statement, they showed that nucleic acids were capable of easily doing what proteins cannot—reproduce themselves. Naturally, it was easy to accept that nucleic acids were the basis, and hence the defining element, of life. This is the reason that some consider viruses alive, despite their dependence on hosts.

However, even this definition has its limit. How do we evaluate red blood cells, which do not have nuclei? Are erythrocytes alive? They must be somehow, because not only do they perform the critical task of transporting oxygen throughout the body; they must also be able to maintain structural integrity while repeatedly being exposed to tremendous shear stresses—not just for seconds or minutes, but for 120 days in humans.

Others attempt to describe life as consisting of eight characteristic processes: movement, excretion, respiration, irritability, growth, reproduction, adaptability, and nutrition. However, you can find exceptional examples of life where one or more of these processes are lacking, rendering this definition unreliable. Besides, this convention defines life using eight complexes processes, each of which is nearly impossible or very difficult to recreate in the laboratory. Clearly, it would not be practical to define the minimal cell on the basis of these processes alone.

2.3 Autopoiesis

The Chilean biologist-philosophers Humberto Maturana and Francisco Varela decided to develop a broader, and hence more versatile, definition of life. They defined life as a system that is autopoietic. Autopoiesis, they proposed, is characterized by (1) a system enclosed by a boundary; and (2) the enclosed system being a self-repairing mechanism.

In cells, this boundary is the lipid bilayer that makes up cell membranes. Cells themselves contain the materials and processes to produce every component of themselves, including the cell membrane. This biological factory is needed in order to replace material lost from the system in the form of waste or wear. The source of this material, or its precursor, is the environment. This means that the boundary has to be selectively permeable, allowing materials for self-construction to enter and allowing waste products to leave. An autopoietic system is, therefore, a factory whose basic function is to repair itself.

Figure 2.3 represents autopoiesis in a very basic form. There is a bounded system at the center, into which the environment provides an input. This input is processed and transformed by the machinery enclosed into materials for reforming parts of the bounded system. This process, in turn, produces an output, which then enters the environment. This basic scheme can describe most living systems, except that each component you see will be different.

Autopoietic theory

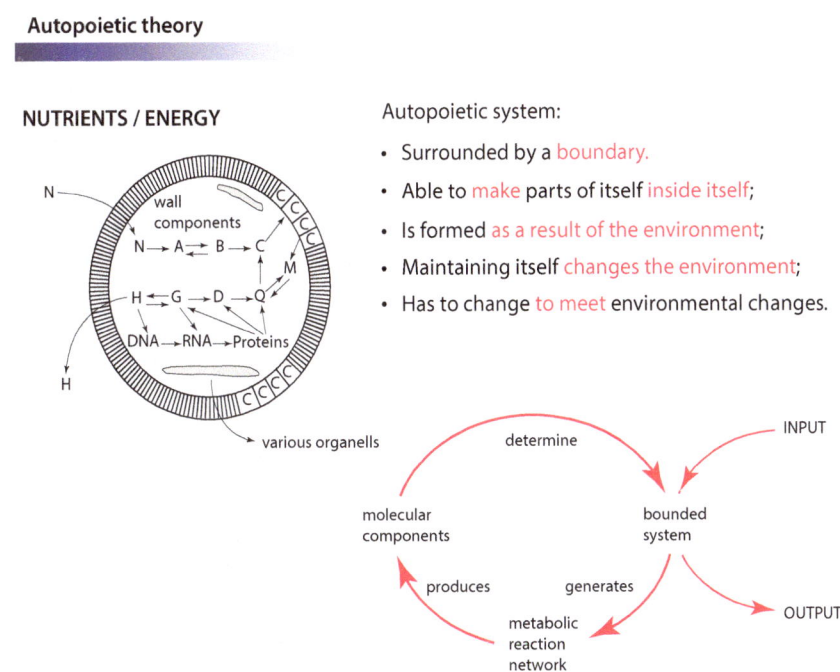

NUTRIENTS / ENERGY

Autopoietic system:

- Surrounded by a boundary.
- Able to make parts of itself inside itself;
- Is formed as a result of the environment;
- Maintaining itself changes the environment;
- Has to change to meet environmental changes.

Fig. 2.3 Some basic features of an autopoietic system. These are systems that are confined by, and includes, a boundary such as a membrane. This boundary separates the interior from the surrounding environment. There is, however, exchange of material between the two spaces. Enclosed within the boundary are the components of a machine that recreates itself, including the boundary, provided it has the necessary building material. Such material enters the system across the boundary while the waste products created by the enclosed machinery are similarly able to escape. In essence, a basic autopoietic system is a self-repairing machine. The figure on the left illustrates this concept using the example of a simplified cell. (Luisi, 2003)

2

■ **Fig. 2.4**   This figure shows
how autopoiesis can accommo-
date other biological definitions
of life. In this example, it shows
how it can similarly describe life
based on the central dogma of
molecular biology (Luisi, 2003)

A broader definition

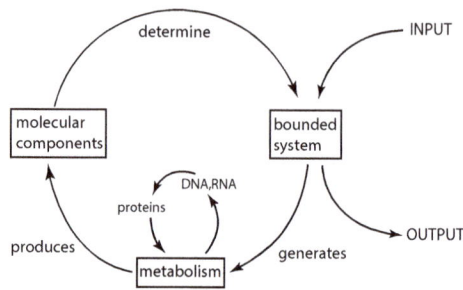

Includes nucleic acid-based definitions

For example, autopoiesis can describe a cell as an enclosed nucleic acid–based system (■ Fig. 2.4). Here, the factory enclosed in the boundary is the molecular biological components of the central dogma of molecular biology: DNA, RNA, and proteins. Here, the DNA produces RNA, which produces proteins, which produce the components of the entire system, using basic materials from—and expelling waste material into—the environment.

However, life is more than a status quo. So far, we have described autopoietic systems repairing themselves and maintaining their structures. Can autopoiesis also account for the growth and death observed in cells? It can if one considers that autopoiesis comprises two net processes: (1) generation of the structure; and (2) degradation of the structure. Since each process comprises complex biochemical reactions, they each have a reaction rate.

Here, simple equations allow us to describe growth, homeostasis, and death in terms of comparative reaction rates (■ Fig. 2.5).

When the rate of generation matches that of degradation, the system is maintaining a status quo. When the rate of generation is greater than that of degradation, the system is growing. In the case of the reverse, the system is dying. Death will occur if the system is not able to cope with changes to the environment that promote degradation.

The environment has a major influence on life. It is the initial conditions of the environment that give rise to a viable autopoietic system. As such, the autopoietic system is naturally able to use the environment to maintain itself. However, by exploiting the environment, the autopoietic system, in turn, perturbs and changes it. Sometimes, these changes are so significant that the autopoietic system also has to change in order to adapt. In this way, the autopoietic system and its environment develop together. This means that every autopoietic system and its environment share an evolutionary history. This is a cyclic process, which goes on until the autopoietic system is no longer able to cope with the new conditions. In other words, autopoietic systems are also capable of evolving.

■ **Fig. 2.5** This figure illustrates how autopoiesis can be described quantitatively. If an autopoietic system is one that is able to maintain itself in the face of constant material gain and material loss, then one can say that its anabolic processes balances out its catabolic processes. If both processes are represented by rate equations, then the rate of anabolism (v_{gen}) would be equal to the rate of catabolism (v_{dec}). When v_{dec} exceeds v_{gen}, the system would be dying. In contrast, the system would be growing if v_{gen} exceeds v_{dec} (Luisi, 2003)

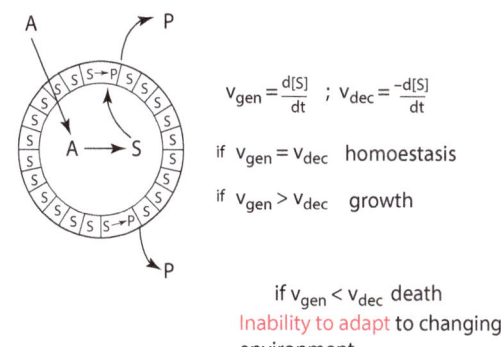

Growth, homeotatis, death

$$v_{gen} = \frac{d[S]}{dt} \quad ; \quad v_{dec} = \frac{-d[S]}{dt}$$

if $v_{gen} = v_{dec}$ homoestasis

if $v_{gen} > v_{dec}$ growth

if $v_{gen} < v_{dec}$ death
Inability to adapt to changing environment

Most importantly, autopoiesis provides us with clear criteria for ascertaining whether something is alive or not. These criteria can be used as engineering objectives to allow us to create autopoietic systems. The simplest, or minimal, forms of such autopoietic systems would serve as artificial minimal cells.

2.4 The Top-Down Approach: The Minimal Bacterial Genome As An Example

With this functional definition of the minimal cell, let us examine how the top-down approach was used to approximate such a system. This was work done at the J. Craig Venter Institute, founded and named after Craig Venter. His research team has been trying to identify the minimal bacterial genome, which would encode a minimal bacterial cell, under specific culture conditions.

There were two major reasons for starting with bacteria. First of all, it seems probable that, being less complex, bacterial cells might require fewer genes for encoding their structure and function. Furthermore, bacterial genomes, at the time when this work was undertaken, were among the first genomes sequenced. In fact, the first such sequencing was performed on *Mycoplasma genitalium.*

The mycoplasmas are a group of bacteria whose genomes are thought to be one of the smallest, with that of *M. genitalium* mistakenly thought to be the smallest of all. It should thus be easier to identify nonessential genes using *M. genitalium.* To do this, the research team compared the genome of *M. genitalium* with that of *Hemophilus influenzae.* They reasoned that both bacterial species must share a common, minimal set of genes that would define them as bacterial cells. All other genes would be supplementary and would code for characters that define their species, or may have other functions unrelated to maintaining life.

From this comparison, they found 250 genes that were common. These 250 genes must, they reasoned, be enough to code for a minimal bacterial cell. They were wrong, in fact. Systematic mutagenesis of these genes indicated that even these 250 included genes not essential to life. In other words, the minimal bacterial genome should actually be even smaller. In this way, they approached identifying the minimal set of genes.

2

2.5 The Bottom-Up Approach: Chemical Autopoiesis

Pietro Luigi Luisi and his work on the minimal cell exemplify the use of the bottom-up approach to approximate the minimal cell. In early experiments, he employed ethyl caprylate, which is hydrolyzed at a high pH into caprylate. Caprylate is amphiphilic and hence, at its critical micelle concentration, is able to self-assemble into micelles (❏ Fig. 2.6).

There is a clearly defined boundary, which is almost a cell membrane bilayer. Nonetheless, it isolates the hydrophobic interior from the hydrophilic environment—one of the basic requirements of an autopoietic system. Furthermore, these micelles can also entrap ethyl caprylate and hydrolyze it to caprylate through autocatalysis. In this way, the micelles are able to produce the very material from which they are made. This is, in essence, a primitive autopoietic system created using raw materials.

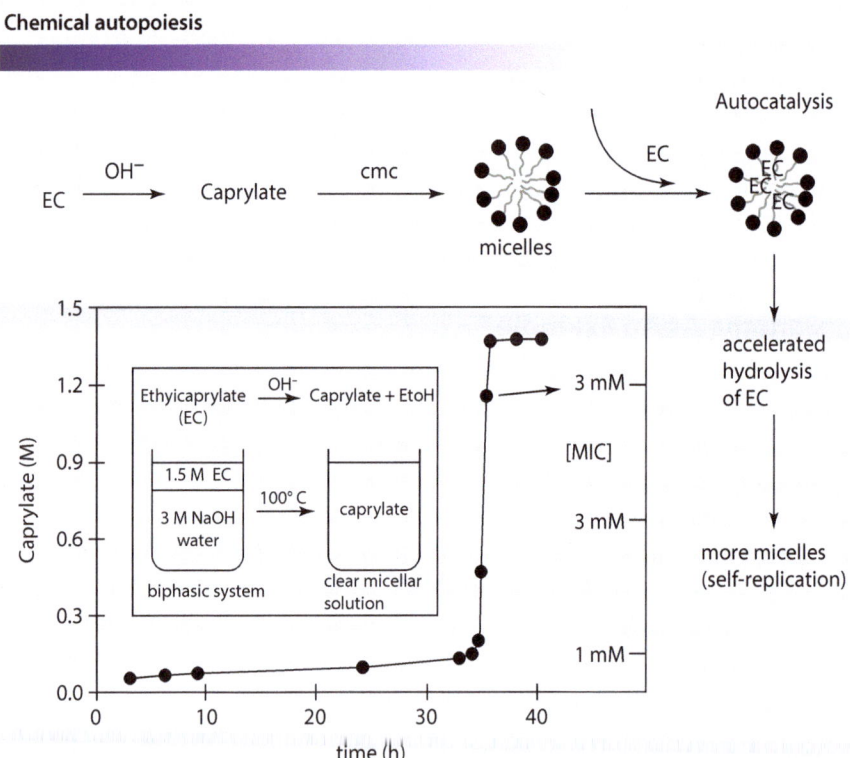

❏ **Fig. 2.6** An example of self-assembling molecules that have characteristics of an autopoietic system. Here, ethyl caprylate (EC) is hydrolyzed by a high pH into caprylate. Caprylate is amphiphilic and self-assemble into micelles which entrap ethyl caprylate. This confinement, itself, catalyzes hydrolysis of ethyl caprylate to caprylate, so contributing to the micellar boundary. In this way, a primitive autopoietic system using raw materials is generated (Luisi, 2003)

However, a micelle is not a membrane bilayer. A bilayer comprises two layers of such amphiphilic material, self-assembled in such a way that the hydrophobic domains face each other and the hydrophilic domains face either side of the membrane. This, too, Luisi has managed to emulate using surfactants. Again, he has been able to select surfactant precursors capable of autocatalysis into the self-assembling amphiphilic forms. The results are bilayered surfactant vesicles that not only can grow but also at a critical size would divide into multiple vesicles. In other words, these surfactant vesicles could not only grow, but reproduce as well! To make the system even closer to that in cells, these surfactants have also been successfully replaced with lipids that behave in a similar manner, except that they formed bilayered membrane vesicles instead. Such work demonstrates how, using only simple molecules to construct more complex structures, we are fast approaching autopoietic systems that resemble actual cells.

2.6 Autopoietic Systems and Their Environment

As mentioned, the state of the environment influences the viability of an autopoietic system. By understanding this relationship, we can understand under which conditions life will arise, under which it will thrive, and which will kill it. Clearly, any massive change to the environment, such as climate change, will present a serious challenge to autopoietic systems. How would life adapt to massive environmental changes, such as climate change?

One is tempted to assume that life is sustainable only under the conditions commonly assumed to be amenable or even critical to life. These include conditions of temperature, oxygen, and moisture. However, it is thought that the evolution of cyanobacteria, about 2.3 billion years ago, led to the sudden accumulation of oxygen in the atmosphere. Initially, this was absorbed by seabed rocks and the ocean mass, but later, it escaped into the atmosphere. Here, together with the greenhouse gas methane, it caused a massive shift in global temperatures. This is described as the Great Oxygenation Event.

Life until then was largely anoxic. The sudden appearance of highly reactive oxygen was too much for most terrestrial life to adjust to. Nonetheless, life was possible in a pre-green earth and would likely still be possible in a postgreen earth. However, it would be a different kind of life. Understanding this reminds us to consider other unconventional conditions under which autopoietic systems might arise and persist, such as conditions created in the laboratory.

This last point is particularly important where the impact of such synthetic organisms on existing life is concerned. One way to ensure that synthetic cells do not overwhelm or replace existing flora or fauna is to ensure they cannot thrive outside the laboratory nor interact with and modify existing life. Creating autopoietic systems that can persist only under non-natural conditions would be one way to control the spread and growth of synthetic organisms. This concept will be re-examined in the last chapter.

> **Take-Home Messages**
>
> 1. The minimal cell describes the simplest possible form of cellular life, under a specific set of conditions.
> 2. A minimal cell can be modified to create more complex cells or cells with special functions.
> 3. This process would allow us to learn how simple cells might evolve into complex ones.
> 4. Researchers use either the top-down or bottom-up approach to try to create the minimal cell.
> 5. For this to be practical, the minimal cell must be defined practically.
> 6. Autopoiesis provides a functional definition of life and, hence, of the minimal cell.
> 7. Autopoiesis is the ability of a membrane-bound system to use material from its environment to produce or repair all parts of itself, including the membrane boundary. In the process, any waste material is released into the environment.
> 8. Autopoiesis is made possible by the environment and, in turn, affects the environment.
> 9. The ability of an autopoietic system to adapt to changes in the environment determines whether it will thrive.
> 10. Creating autopoietic systems that can thrive only in non-natural environments is a means to reduce any undesirable impact such synthetic organisms would have on existing life.

Further Reading

Caschera F, Stano P, Luisi PL (2010) Reactivity and fusion between cationic vesicles and fatty acid anionic vesicles. J Colloid Interface Sci 345:561–565

Chiarabelli C, Stano P, Luisi PL (2009) Chemical approaches to synthetic biology. Curr Opin Biotechnol 20:492–497

Fraser CM, Gocayne JD, White O, Adams MD, Clayton RA, Fleischmann RD, Bult CJ, Kerlavage AR, Sutton G, Kelley JM et al (1995) The minimal gene complement of *Mycoplasma genitalium*. Science 270:397–403

Gibson DG, Glass JI, Lartigue C, Noskov VN, Chuang RY, Algire MA, Benders GA, Montague MG, Ma L, Moodie MM et al (2010) Creation of a bacterial cell controlled by a chemically synthesized genome. Science 329:52–56

Holland HD (2006) The oxygenation of the atmosphere and oceans. Philos Trans R Soc Lond B Biol Sci 361:903–915

Hutchison CA 3rd, Chuang RY, Noskov VN, Assad-Garcia N, Deerinck TJ, Ellisman MH, Gill J, Kannan K, Karas BJ, Ma L et al (2016) Design and synthesis of a minimal bacterial genome. Science 351:aad6253

Luisi PL (2003) Autopoiesis: a review and a reappraisal. Naturwissenschaften 90:49–59

Stano P, Luisi PL (2013) Semi-synthetic minimal cells: origin and recent developments. Curr Opin Biotechnol 24:633–638

Watson JD, Crick FH (1953) Molecular structure of nucleic acids; a structure for deoxyribose nucleic acid. Nature 171:737–738

Synthetic Proteins

© Springer International Publishing AG 2018
E.-K. Ehmoser-Sinner, C.-W. D. Tan, *Lessons on Synthetic Bioarchitectures*, Learning Materials
in Biosciences, https://doi.org/10.1007/978-3-319-73123-0_3

What You Will Learn in This Chapter
"Toward a Synthetic Genome" Section
We will now look at an example of human-designed proteins whose natural counterparts have never been observed. We will show how even nearly random protein sequences can have surprising biological functions. We will explain how the novelty of such proteins reflects the novelty of the genes encoding them, as well as how a minimal set of such genes capable of generating autopoiesis would constitute a novel and truly synthetic genome.

3

3.1 Synthetic Proteins: What Are They?

"Synthetic protein" is still not a precise terminology; it can describe a protein synthesized from living cells (e.g., harvested from a biofermenter), a hybrid macromolecule (e.g., a fluorescently labeled antibody), or even a synthetic alternative to a protein with a desired functionality (e.g., an enzyme), which can be based on completely synthetic compounds, synthesized in the laboratory. Let us start with the aim of such material development first: why bother with synthetic proteins when the proteins built by nature are obviously working quite well on our planet?

Such attempts are often related to pharmaceutical approaches, e.g., synthetic proteins as novel drugs, such as novel antibiotics, novel implant materials, and oncotargets. Antimicrobial peptides, for example, are on the way to revolutionizing the view on drug development, as inherent peptidic fragments of human albumin, for example, play a role in the context of virus infections and, moreover, in diabetes and Alzheimer disease, as recent findings have indicated.

Such molecules are targets for stabilizing and transformation endeavors in order to perform "communication" tasks in misregulatory contexts. This would be a major breakthrough and brings us back to the introduction of this book (see ▶ Chap. 1) regarding synthetic bioarchitectures as a converging field of research: communication with nature on the level of molecules.

Why is the molecular level of such interest and potential? It is not only to "understand" and "define" compounds; the atomistic/molecular level is simply the common denominator of life—whether a protein derives from humans or plants is irrelevant in the moment of its presence in a suitable environment. This is the prerequisite for biotechnology: when bacterial species produce active compounds based on the genetic code of humans—for example, human insulin, synthesized in bacterial species. Another example to demonstrate the potential of biotechnology in general is the possibility (and the ethical issue) of crossing the borders of theoretically all living species. The gene coding for an autofluorescent protein from *Photobacterium leiognathi* has been transferred into the higher plant specimen *Nicotiana tabacum*, in which it performs its function even it is translated in a very different biochemical and genetic context (Krichevsky et al., *PLoS One*, 2010, 5: Issue 11, e15461) (◘ Fig. 3.1).

The context of molecular biology offers a broad spectrum in employment of proteins for therapeutic use. Genetic manipulation strategies enable the synthesis of precise biological compounds, which are partly modified according to their desired function (optimized in specificity) or even combined with synthetic (e.g., carrier) materials.

And, of course, the race to obtain completely novel functionalities in designing proteins de novo has already started!

In the attempt to "build" synthetic proteins de novo, an interesting example demonstrates the capability of such autologous synthetic proteins.

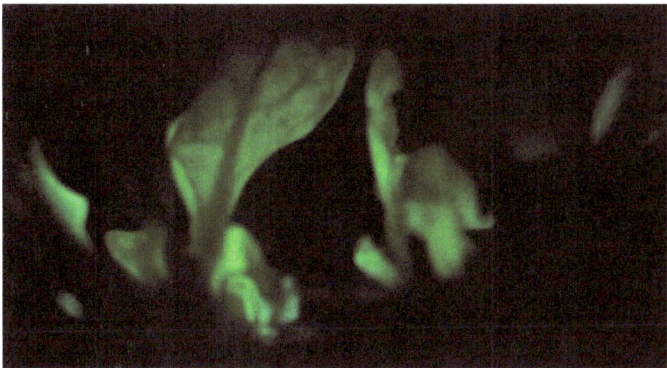

 Fig. 3.1 Visual detection of autoluminescence in LUX-TrnI/TrnA plants. **a** Photograph taken in the dark with a handheld consumer camera (Nikon D200; AF-S Micro Nikkor 105.0 mm 1:2.8 G ED lens; exposures 5 min at f/4.5, 105 mm focal length, ISO 3200). **b** Photographs of transplastomic and wild-type plants taken with lights on or off (Krichevsky et al., *PLoS One*, ▶ https://doi.org/10.1371/journal. pone.0015461.g005)

3.2 Toward a Synthetic Genome

In ▶ Chap. 2, we examined how the J. Craig Venter Institute (JCVI) attempted to determine the minimal set of genes encoding bacterial life. This work went on to encourage attempts at the institute to construct a synthetic bacterial genome, using raw materials. Although the JCVI's work resulted in a synthetically assembled genome, the product was not bona fide synthetic, since the proteins encoded were of existing types. A truly synthetic genome is novel not because of the way it is constructed but in the proteins it encodes. As such, a truly synthetic genome should encode proteins that have never been observed in our biosphere. This has not been achieved yet.

We present the work of Fisher et al. as an example of how this might be possible in the future. In their report, they attempted to show that novel synthetic proteins could be biologically functional. To do this, they first generated 1.5×10^6 partially random 102-residue protein sequences. The only condition they set was that the sequences had to be able to form stable globular structures (▪ Fig. 3.2). This was based on the assumption that a tertiary structure was necessary for most biological functions.

Plasmids encoding each synthetic protein were then transformed into 27 auxotrophs, which can only grow in rich media. Some of the synthetic proteins were able to rescue four of the 27 mutants (▪ Fig. 3.3). Where the auxotrophs were transformed with plasmids encoding *lacZ*, no colonies formed in minimal media. In contrast, rescued mutants were able to form colonies.

Each mutation was rescued by a different set of synthetic proteins. Although each rescue allowed the host to grow in minimal media, the rescues tended to grow more slowly than wild-type hosts (▪ Fig. 3.4). This could be because, unlike wild-type proteins, the synthetic proteins had not been optimized for function through evolution. As such, they might have been selected against in nature.

In trying to understand what functions these synthetic proteins might have, the researchers looked at the mutations that were rescued. Each mutation was in a gene encoding an enzyme critical for the survival of the host in a minimal medium (▪ Fig. 3.5).

The synthetic proteins, despite being completely artificial, could somehow complement the function of these missing enzymes.

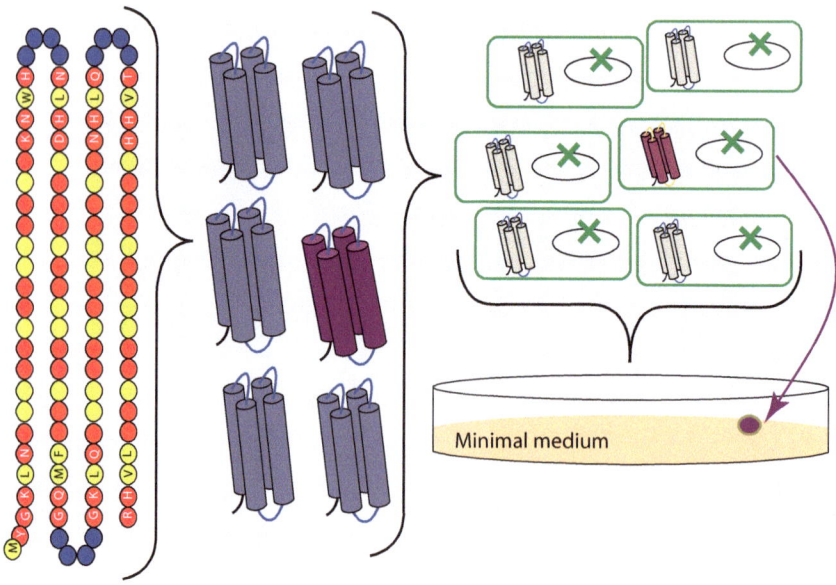

■ **Fig. 3.2**   Design of a collection of novel proteins and rescue of Escherichia coli auxotrophs (Fisher et al. 2011)

Complementation frequency and time required for colony formation on selective media (M9-glucose).

	Complementation Frequency	Time to Grow (Days)
Δ*serB*	~1/50,000	4
Δ*gltA*	~1/1,000,000	4
Δ*ilvA*	~1/3,000	3
Δ*fes*	~1/10,000	2

27 strains that grow in rich media (LB) on minimal media (M9 minimal glucose).

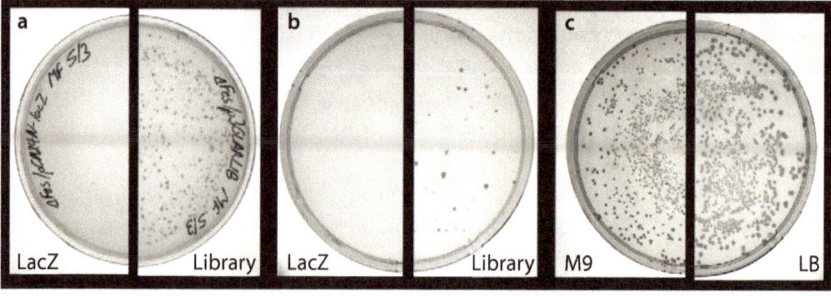

LacZ Library	LacZ Library	M9 LB
Δ*fes* on Minimal	Δ*ilvA* on Minimal	Retransform Syn-IlvA1 into Δ*ilvA*

■ **Fig. 3.3**   Plasmids encoding each synthetic protein were transformed into 27 auxotrophs, which can only grow on rich media. Some of the synthetic proteins were able to rescue four of the 27 mutants (Δ*serB*, Δ*gltA*, Δ*ilvA*, Δ*fes*). Rescued mutants were able to form colonies. This demonstrates how even partially, randomly designed proteins can have effective biological functions

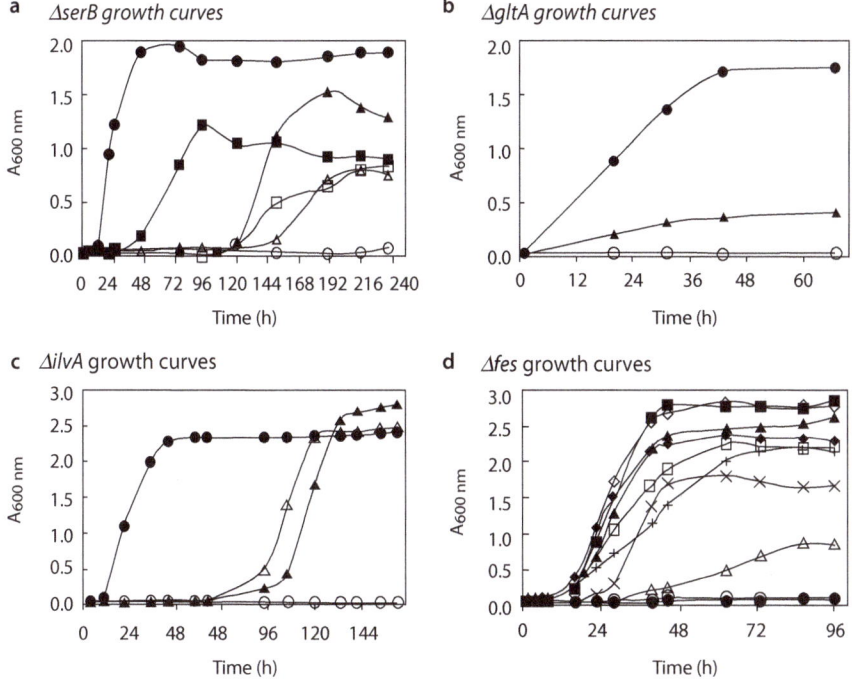

■ **Fig. 3.4** Growth of auxotrophic strains of Escherichia coli in selective liquid media (Fisher et al. 2011)

- *serB*
  - phosphoserine phosphatase
  - responsible for the final step in serine biosynthesis
- *gltA*
  - citrate synthase
  - catalyzes an early step in glutamate biosynthesis:
- *ilvA*
  - threonine deaminase
  - catalyzes the first step in the production of isoleucine from threonine
- *fes*
  - enterobactin esterase
  - enables cells to aquire iron in iron-limited environments
    (Over-expression is toxic.)
                Survival under starvation!

■ **Fig. 3.5** The figure details the product and function of the four mutated genes which the synthetic peptides had rescued. Interestingly, the four mutations were in genes encoding metabolic enzymes

It is still not clear how the synthetic proteins actually rescued the mutants. The following possibilities were suggested: that the synthetic proteins (1) perform the function of the missing enzymes; (2) allow the host to bypass the compromised metabolic pathway; or (3) are not, themselves, complementary but trigger the expression of complementary genes.

3

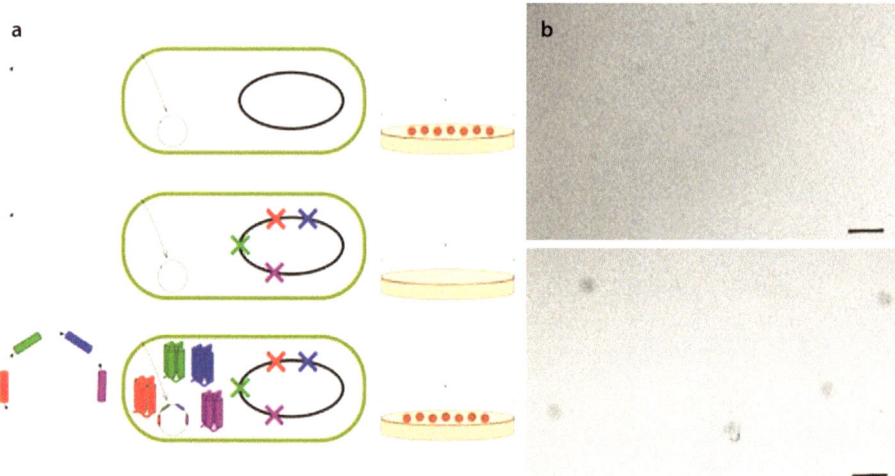

Novel sequences bearing no similarity to naturally evolved proteins
can compensate for deletion of a portion of the E.coli chromosome.

■ **Fig. 3.6**   The synthetic proteins were able to rescue auxotrophs carrying all four of the metabolic enzyme mutations. This demonstrates how multiple synthetic proteins can partially replace the function of portions of a genome, suggesting that an entire genome might be similarly replaced, or designed, to produce a viable system

The experiment was modified to include auxotrophs that carry all four of the rescuable mutations (■ Fig. 3.6). This mutant was then transformed with four of the complementary synthetic proteins. The synthetic proteins were able to rescue the auxotrophs, despite the fact that nearly 0.1% of the host genome had been compromised by the mutations. Might it be possible to similarly replace the remaining 0.9% of the host genome?

This work demonstrates that existing genes and proteins may not constitute all that is functional in existing living systems. Genes encoding novel proteins may be able to address existing needs and may even confer novel functions on their hosts without compromising their viability. As we approach a minimal library of synthetic proteins capable of sustaining life, we approach the emergence of a truly synthetic genome.

■ **Conclusion**

The term "synthetic proteins" describes amino acid–derived compounds that are (1) synthesized by a species under defined conditions (e.g., human insulin synthesized by bacteria); (2) manipulated with additional materials; or (3) derived from de novo design concepts based on the conventional amino acid material context or from noncanonical or even synthetic materials.

┌─ **Take-Home Messages** ───────────────────────────────

"Toward a Synthetic Genome" Section

1. It is argued that a truly synthetic genome should be one that encodes synthetic proteins, not just one that is constructed by humans.
2. It is argued that for proteins to have a biological function, they require tertiary structures at least as complex as stable globular forms.
3. It is possible for specific, de novo globular proteins, comprising partially randomly generated sequences, to rescue conditional mutants.
4. Such synthetic proteins are not as effective as their natural counterparts, suggesting that they might have been selected against in nature, hence their scarcity or novelty.
5. A genome comprising genes that encode only synthetic proteins, and that would give rise to an autopoietic system, might be considered a truly synthetic one.

Further Reading

Fisher MA, McKinley KL, Bradley LH, Viola SR, Hecht MH (2011) De novo designed proteins from a library of artificial sequences function in Escherichia coli and enable cell growth. PLoS One 6:e15364

Ow DW, Wood KV, Deluca M, Dewet JR, Helinski DR, Howell SH (1986) Transient and stable expression of the firefly luciferase gene in plant-cells and transgenic plants. Science 234:856–859

Biomimicry: The Bottom-Up Approach

© Springer International Publishing AG 2018
E.-K. Ehmoser-Sinner, C.-W. D. Tan, *Lessons on Synthetic Bioarchitectures*, Learning Materials in Biosciences, https://doi.org/10.1007/978-3-319-73123-0_4

4

What You Will Learn in This Chapter

1. Model systems used for analysis via synthesis.
2. Principles of engineering biomolecules.
3. Understanding of biomolecules; alchemy to urea again.
4. Membrane mimetics.
5. Membrane components.
6. Understanding of the concept of synthesis from first principles.
7. Understanding of why this is done.
8. Understanding of the various ways this principle is applied.
9. Awareness of examples.
10. The student should be able to explain how the bottom-up approach is applied to synthetic biology.
11. He or she should also be able to explain the uses of this approach.
12. The student should be able to cite and illustrate this use, with examples.

In the search for the minimal cell, it is expected that an entire cell might be difficult to model using current technology. We explain how the bottom-up approach is used to model specific parts, or functions, of cells instead. Such model structures include the cell membrane, proteins, and even the genetic material. Each system is constructed using basic materials, which are rationally combined to form functional structures. Some of these systems might even use materials that are different from their counterparts in the biosphere. Finally, we introduce the idea of synthetic organisms comprising or using noncanonical materials as a safeguard against their accidental release into the environment.

4.1 Synthetic Bioarchitecture: The Bottom-Up Approach

The bottom-up approach thrives on the idea of self-assembly. In the light of self-organizing structures, the dimension of life leans on self-organization. To adapts this concept is only a natural consequence in research applications, dealing with the dimension of molecules—otherwise, it is quite tedious (if not impossible) to manufacture molecular assemblies with mechanical tools. In the 1980s, the famous physicist Richard Feynman stated, "What I cannot create, I do not understand." This is still valid for physics and for biology; this motto is still far in the future for all "living" species. Even Craig Venter, the godfather of "synthetic life," who has "implanted" a functional genome in a living species, has not "created" life from the bottom up. Creating life has always been a dream of mankind, and synthetic bioarchitecture is a realistic view of how far and possibly impossible this goal is, as life might not be encoded in a structure; there may be something more to it.

Let us give a real-world, state-of-the-art example of a bottom-up approach in synthetic biology.

The concept of DNA origami has been developed over the past decade. Of course, DNA does not consist of just four monomers; it also provides unusual charge density and, as a consequence, an inherent and precise spatial control.

- Can we build functional (molecular) Hybrid structures employing the blueprints of Nature?
- Does Nature (living cells), understand' such hybrid architectures?

■ Fig. 4.1   Gustave Doré - The Monkey and the Dolphin (1867), a scene from the book "Fontaine's Fables". This example is intended to illustrate the idea of functional assembly - quasi 'incompatible species', monkey and dolphin, merge into a functional unit

With this famous example, let us start in the world of bottom-up approaches with a concept aiming for employment of functional building blocks and achievement of self-sustaining (or even autopoietic) systems. In the end, it is a protocell, which has been sketched by many but is still unattained, as confinement of the building blocks of life in small spherical objects is still hampered by lack of control of energy influx/efflux and, as a consequence, it is still unable to sustain itself in laboratory conditions. Interestingly, the "end of the game" is the status of equilibrium. At this point, the far end goal is understanding "far from equilibrium reactions with the intention to build in such concepts in bottom-up approaches". If we understand how to combine macromolecules in functional assemblies, we might achieve "sustaining" objects with even unconventional abilities, as Gustave Doré depicted in his graphics of an "impossible" hybrid creature, such as the dolphin in ■ Fig. 4.1, carrying a dog's head and being guided by a monkey. This image depicts the gap to be closed in order to preserve function, while being composed of impossible matches from the evolutionary point of view. How this might go, along with unintended consequences, is discussed in this book under safety issues related to the field of synthetic bioarchitectures (see ▶ Chap. 6).

In this book, we focus on membranous interfaces as a relevant and eligible example of a highly organized, self-assembled, functional structure from nature. We use the concept of bottom-up approaches, as these structures inherently consist of lipid molecules, organized as a two-dimensional crystal structure in a liquid ordered state. We can approach a self-organized spherical object with our available methods and tools for characterizing embedded proteins and even mimetics of simple biochemical feedback cycles (■ Fig. 4.2).

4

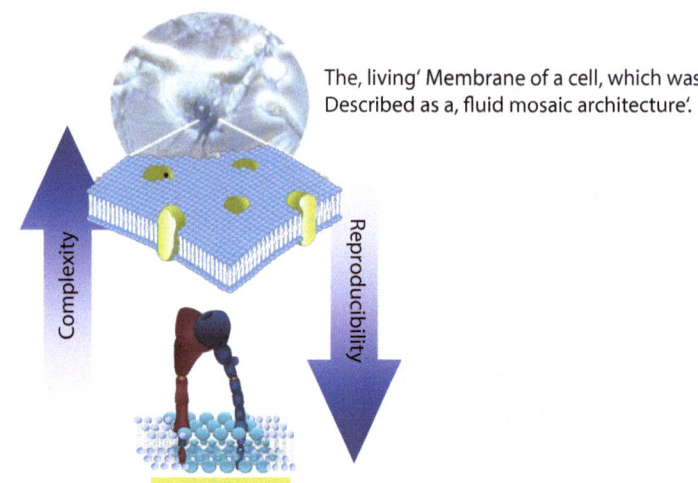

The "relevance" of model systems in science…

The, living' Membrane of a cell, which was Described as a, fluid mosaic architecture'.

Complexity

Reproducibility

A (gold surface tethered) Membrane Model System depicting a polymeric/phospholipid membrane hybrid with an embedded membrane protein: Integrin. (Zaba et al. 2015, ChemBioChem)

Fig. 4.2 We will always have to compromise with the "robustness, e.g., reproducibility" offered by a model system, reflecting only some properties of the naturally integrated system. Cells, as natural locations of interesting biomolecules, such as membrane proteins, are subtle units with complex, inter-calating biochemical responses and the inherent feature of aging in the course of an experiment. On the other hand, we have to start somewhere with the attempt to "catch" a glimpse of biomaterials; as such, we "freeze" objects, such as membrane proteins, in a tethered, metal-anchored, planar lipid membrane architecture in order to understand, for example, ligand receptor interactions

4.2 The Minimal Cell Revisited

As explained in ▶ Chap. 2, synthetic biology employs two basic approaches in its work: the top-down approach and the bottom-up approach. Where these two techniques meet, at their extremes, is the minimal cell. The concept of the minimal cell posits that all cells are advanced variations of a much simpler cell.

An analogy would be a car that consists of just a frame, an engine, axles, and wheels. With this simple chassis, you can then modify it to make a luxury car, an ambulance, or even something novel. The idea of the minimal cell is similar. This method of stripping down to the minimum is the top-down approach.

Unlike the top-down approach, the bottom-up approach attempts to put together diverse building materials to construct complex structures. In the process of trying to recreate parts of (or whole) cells, we will hopefully learn how each part evolved and how each improvement solves a problem (◻ Fig. 4.3). By observing what exists in nature, we hope to find solutions to engineering problems.

Applied to Synthetic Biology

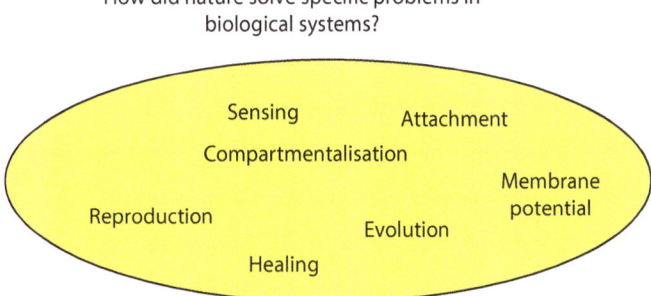

"What I cannot create,
I do not understand."

"Know how to solve every problem
that has been solved."

How did nature solve specific problems in
biological systems?

Sensing   Attachment

Compartmentalisation

Membrane
potential

Reproduction   Evolution

Healing

Can we also use these solutions?

□ Fig. 4.3 Studying, reverse engineering and mimicking various structures and processes in living systems can teach us how to develop solutions to complex biological problems. The figure quotes from Richard P. Feynman

4.3 The Reductionist Approach

However, a cell is more than just a structure. We also have to recreate some of the processes that occur in it (□ Fig. 4.4). As such, attempting to construct whole cells from scratch is not trivial. A more feasible approach would be to start by making parts of a living cell, such as the cell membrane.

Cell membranes comprise (1) various lipids, which form the bilayer; (2) proteins, which are embedded in or attached to them; (3) carbohydrates, which modify both; and (4) cholesterol, which modulates the membrane fluidity (□ Fig. 4.5). Using combinations of lipids and cholesterol, one can produce artificial membranes of various forms.

However, such membranes would lack the critical membrane proteins. Various methods exist to synthesize proteins and incorporate them into artificial membranes. One powerful approach is to supplement cell-free protein synthesis reaction mixes with artificial membranes (□ Fig. 4.6). During protein production, membrane proteins would integrate into the artificial membranes.

With artificial membrane vesicles, one could mimic the surface structure of bacteria. Most bacteria are coated with surface-layer (S-layer) proteins. They are found to form regular patterns on the surface membranes of both Gram-positive and Gram-negative bacteria, and serve various functions including surface attachment and virulence. One can mimic this surface by preparing pure S-layer proteins and, under the right conditions, allowing them to crystallize on the surface of artificial liposomes (□ Fig. 4.7).

4

Not quite the minimal cell……

Mimicking just the parts?
- Cell surface membrane
- Nucleus
- Cell wall
- etc

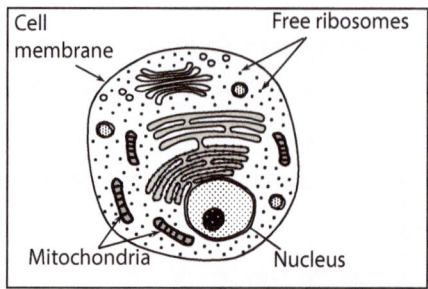

Mimicking just the processes?
- Information transfer
- Mass transfer
- Photosensing
- etc

Fig. 4.4   Do we mimic biological structures or biological processes?

Mimicking structures

Cell membranes

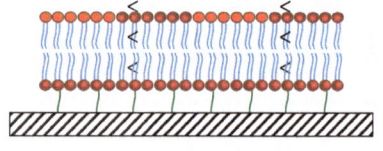

Artificial membranes

Lipids
- Phospholipids
- Sphingolipids
- Glycolipids

Cholesterol
Proteins
Carbohydrates

Tethered membranes

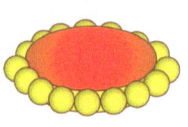

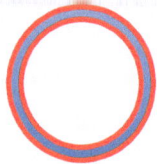

Nanodics   Liposomes

Fig. 4.5   Examples of artificial lipid membranes. Membranes can be constructed from some, or all, of the basic components found in natural cell membranes. Different membrane structures such as tethered planes and liposomes allow synthetic membranes to be used in diverse ways

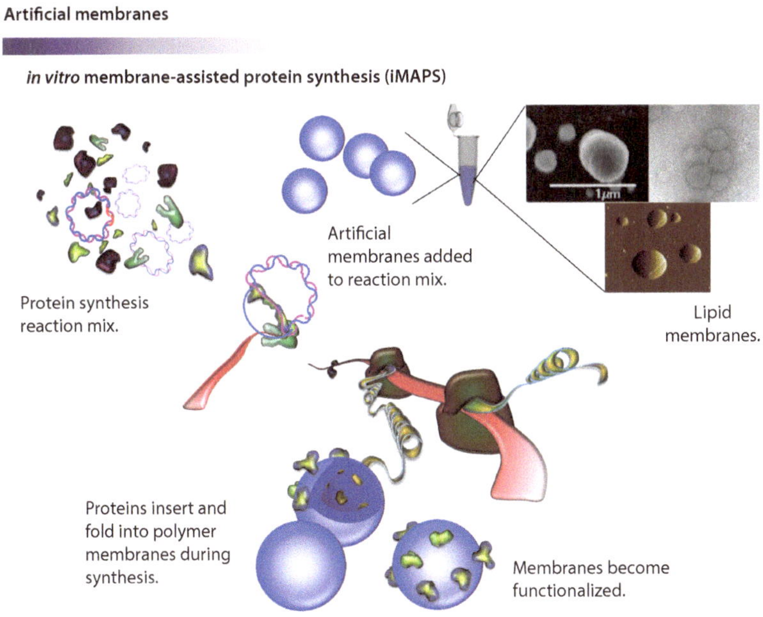

Artificial membranes

in vitro membrane-assisted protein synthesis (iMAPS)

Protein synthesis
reaction mix.

Artificial
membranes added
to reaction mix.

Lipid
membranes.

Proteins insert and
fold into polymer
membranes during
synthesis.

Membranes become
functionalized.

□ **Fig. 4.6** The figure illustrates *in vitro* membrane-assisted protein synthesis. Artificial membranes are added to cell-free protein synthesis reaction mixes. As nascent proteins are produced, they are though to insert into, fold and orient themselves in, the membranes

Mimicking structures

Mimicking bacterial surfaces

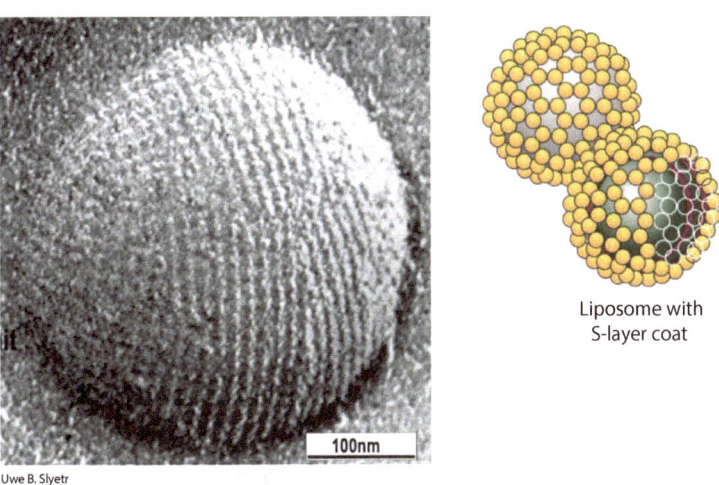

100nm

Uwe B. Slyetr

Liposome with
S-layer coat

□ **Fig. 4.7** An example of how vesicles might be modified with a coat of Surface-layer (S-layer) proteins, in order to mimic bacterial cell surfaces. (Sleytr)

4

Silicon is found in living cells!

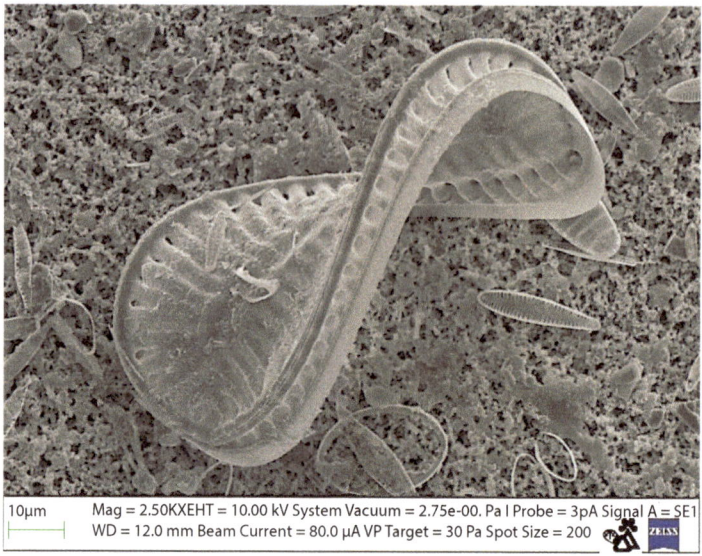

| 10µm | Mag = 2.50KXEHT = 10.00 kV System Vacuum = 2.75e-00. Pa I Probe = 3pA Signal A = SE1 |
| | WD = 12.0 mm Beam Current = 80.0 µA VP Target = 30 Pa Spot Size = 200 |

Surirella spiralis **diatom**

■ **Fig. 4.8**   An example of how, despite being associated with all things artificial in modern culture, silicon is often incorporated into the structure of living systems. (Angeli, 2016)

4.4 **Considering Materials**

In the attempt to mimic natural systems in terms of structure and function, researchers are free to use synthetic materials. This provides the advantage of a wider choice of materials, as well as versatility. As such, one can consider the use of atypical materials, such as silicon. Although it is associated with synthetic products, silicon can actually be found in living systems.

Silicon belongs to the same group of elements as carbon, and hence can form many molecules similar to carbon dioxide and methane. Like carbon, silicon is capable of concatenation. As such, some have posited that life based on silicon might also be possible. An early pioneer of this idea was the nineteenth-century German astrophysicist Julius Scheiner.

Many terrestrial lifeforms, such as diatoms, bacillariophytes, and sponges, use the silicon derivative silaffin as a building block (■ Fig. 4.8). In diatoms, they form exoskeletons with distinct shapes and properties. Instead of S-layer proteins, one can think of coating artificial liposomes with such material.

Polymer

<u>Sample Name</u>: **Poly(butadiene-b-ethylene oxide)**
Poly butadiene rich in 1,2 or 1,4 microstructure

<u>Sample #</u>: **P9089-BdEO**
(Poly butadiene rich in 1, 2 microstructure)

Structure of 1,2-rich microstructure:

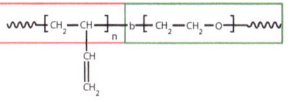

Structure of 1,4-rich microstructure:

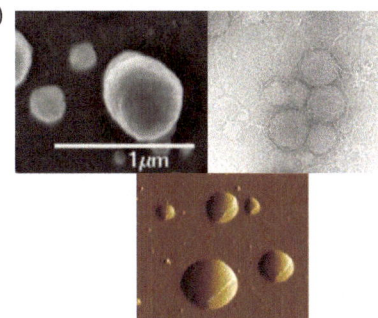

<u>Sample Name:</u> **Poly(ethylene oxide-b-dimethyl siloxane-b-ethylene oxide)**
<u>Sample #:</u> **P7300-EODMSEO**
Structure:

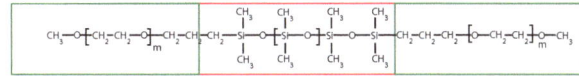

■ **Fig. 4.9** Examples of amphiphilic polymers. Just like amphiphilic lipids, polymers with hydrophilic (*red box*) and hydrophobic (*green box*) domains may also self-assemble into membrane bilayers. Tri-block copolymers comprising hydrophilic domains flanking a hydrophobic domain may even form monolayers which would still yield the hydrophilic-hydrophobic-hydrophilic internal character typical of biological membranes

4.5 Unconventional Materials

One might even replace the lipids of liposomes with synthetic amphiphilic polymers, such as poly(butadiene)-poly(ethylene oxide) (■ Fig. 4.9). In aqueous environments, these are also able to self-assemble into membrane bilayers. Such membranes would be more stable against oxidation and mechanical force than liposomes.

Even fundamental cellular functions, such as gene expression, are amenable to mimicry using synthetic material. An example is xDNA, which is DNA comprising bases extended with an additional benzene ring, in addition to the natural four (■ Fig. 4.10). These extended bases are fluorescent and can base pair with normal nucleotides. Their use is expected to teach us more about the behavior of natural DNA.

Another example is the expanded genetic code. Here, stop or nonsense codons in bacteria are used to encode nonstandard amino acids (■ Fig. 4.11). To do this, a mutant

4

Mimicking materials

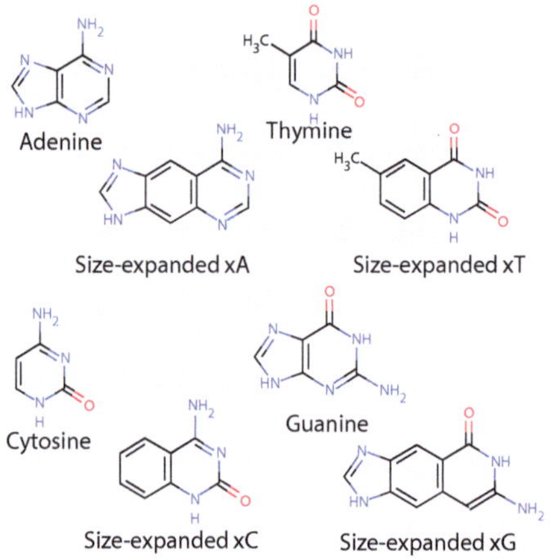

Adenine

Thymine

Size-expanded xA

Size-expanded xT

Cytosine

Guanine

Size-expanded xC

Size-expanded xG

- Extended bases are fluorescent.

- Can base-pair with normal nucleotides as per rules of base-paring.

- xDNA is wider than natural DNA by a benzene ring's width and has a longer pitch.

- Teach us more about the behavior of natural DNA.

■ **Fig. 4.10** Unlike their natural counterparts, size-expanded nucleic acids are fluorescent. Despite the addition of a benzene moiety, each is able to base-pair in the same way as their natural counterparts. (Lynch et al., 2006)

Mimicking materials

Expanded genetic code

- Stop or nonsense codons in bacteria are hijacked to encode non-standard amino acids.

- Non-standard amino acids:
  - Fluorescent amino acids
  - Functionalised amino acids
  - etc

- Allow us to probe protein function with greater versatility.

- Will not interfere with the existing molecular biology.

- Will not be able to survive outside of the host.

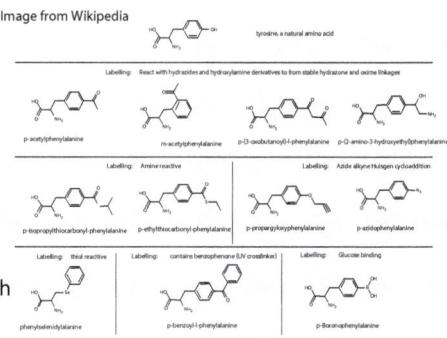

Image from Wikipedia

Hijacked codon
(e.g. amber codon, UAG)
Mutant tRNA
Mutant aminoacyl tRNA
synthetase

■ **Fig. 4.11** Transfer RNA (tRNA) that complement stop or redundant codons can be harnessed to encode amino acids not among the 20 universal protein-forming ones. For this to work, a mutant amino acyl-tRNA synthetase must also be found which is capable of activating the chosen tRNA with the non-standard amino acid

transfer RNA (tRNA), which will only base pair with a stop or nonsense codon, is activated with a fluorescent or chemically reactive amino acid. Using them for gene expression will allow us to probe protein function with greater versatility.

Each of these endeavors will teach us basic principles of engineering, derived from observing how natural systems have evolved to adjust to obstacles. Eventually, the mimicry of biological systems might employ purely synthetic materials and perhaps give rise to novel biological functions. Finally, in the last chapter, we will see how this strategy might be exploited as a safety measure against the accidental release of synthetic organisms into the natural environment.

Take-Home Messages

1. The bottom-up approach attempts to use basic materials to construct complex structures mimicking either the structure or the function of living systems.
2. As whole cells are too complex to mimic entirely, some researchers attempt to mimic only specific structures, such as the plasma membranes of prokaryotic and eukaryotic cells.
3. These attempts might also include the use of unconventional materials for building such structures.
4. These materials include functioning substitutes for amino acids, nucleic acids, and phospholipids, among others.
5. Such synthetic characters might serve as a safeguard against the accidental release of synthetic organisms.

Further Reading

Krueger AT, Peterson LW, Chelliserry J, Kleinbaum DJ, Kool ET (2011) Encoding phenotype in bacteria with an alternative genetic set. J Am Chem Soc 133:18447–18451

Lynch SR, Liu H, Gao J, Kool ET (2006) Toward a designed, functioning genetic system with expanded-size base pairs: solution structure of the eight-base xDNA double helix. J Am Chem Soc 128:14704–14711

Nallani M, Andreasson-Ochsner M, Tan CW, Sinner EK, Wisantoso Y, Geifman-Shochat S, Hunziker W (2011) Proteopolymersomes: in vitro production of a membrane protein in polymersome membranes. Biointerphases 6:153–157

Pamirsky IE, Golokhvast KS (2013) Silaffins of diatoms: from applied biotechnology to biomedicine. Mar Drugs 11:3155–3167

Sumper MK, N. (2004) Silica formation in diatoms: the function of long-chain polyamines and silaffins. J Mat Chem 14:2059–2065

Tan DC, Wijaya IP, Andreasson-Ochsner M, Vasina EN, Nallani M, Hunziker W, Sinner EK (2012) A novel microfluidics-based method for probing weak protein-protein interactions. Lab Chip 12:2726–2735

Xie J, Schultz PG (2005) Adding amino acids to the genetic repertoire. Curr Opin Chem Biol 9:548–554

The Tools

© Springer International Publishing AG 2018
E.-K. Ehmoser-Sinner, C.-W. D. Tan, *Lessons on Synthetic Bioarchitectures*, Learning Materials
in Biosciences, https://doi.org/10.1007/978-3-319-73123-0_5

What You Will Learn in This Chapter

The ambitious nature of projects in synthetic biology requires special methods to match them. For those projects that require extensive modification of genes and others perhaps whole genomes, conventional techniques used in molecular biology have to be improved to meet their needs. These needs include the necessity for gene modification methods to be reliable, easy to handle, and compatible between laboratories. This implies the need for certain standards, particularly for the materials and methods used. One approach to achieve this is to reduce the genetic material manipulated to highly interchangeable and interconnectable modules. We will look at how BioBricks allow us to do this. Another requirement is that the techniques used be precise and capable of large-scale changes to the target genetic material. We will see how the CRISPR/Cas9 system was developed to meet this need, as well as the range of DNA modifications it is capable of. Finally, the hazards posed by both technologies, as well as how those concerned have chosen to deal with them, are discussed.

5.1 Modularity and Standardization

As we have seen, synthetic biology hopes to play a significant role in the pharmaceutical industry. There are attempts to create novel antibiotics to tackle the challenge of rising resistance, new kinds of implants that would be more biocompatible, and new ways to target drugs more effectively, among other projects.

However, a technology's quality has to be taken to a higher level whenever it reaches industry. This is for reasons of safety, reliability and, very importantly, compatibility. When whole systems have to be redesigned to produce a new but related product, significant resources are wasted. Compatibility reduces this problem. Consider the Luer taper standard. As long as components contain the standard Luer male part, they can fit any other component that carries the Luer female part. This allows us to mix and match different components, easily giving us a wide range of solutions to a problem.

Synthetic biology tries to address this issue by applying principles of engineering such as abstraction and modulation. Complex systems are gradually built up from simpler systems and these from simpler parts. At each step, the materials, products, and processes have to meet rigorous standards of quality and reproducibility.

5.2 Reliability and Compatibility in Molecule Biology

Synthetic biology began with, and is still largely concerned with, creating novel biological functions in existing organisms. To do this rationally, one needs to use an engineering approach. Very simply put, this involves identifying a problem, designing a solution, modeling the solution mathematically or on a small scale, testing the solution, and identifying any new problems. This is iterated until the original problem is solved. One way to make this process more efficient is to build in certain levels of reliability so that certain tests of quality need not be repeated.

5.3 Establishing Standards

Consider this problem in the case of molecular biology. This technique is critical to those synthetic biology projects that rely heavily on manipulating genes and gene expression, such as pathway engineering. To make such endeavors more reproducible and, generally, less difficult for both veterans and novices alike, it would be useful if standardized parts, tools, and processes were available.

In this case, our products are vectors and the parts in question are DNA fragments. Each fragment would have a specific function, such as a promoter, ribosome binding site, forward transcription start signal, stop signal, and so on. Our tools would be enzymes and host cells. Our processes would include DNA restriction, ligation, transformation, host selection, and so on.

These are, of course, established techniques in molecular biology. However, it is often the case that one lab uses an entirely different set of materials, tools, and processes from another. An example of how this diversity affects compatibility between collaborating laboratories is that often a vector supplied by one partner is not compatible with a downstream process used by the other. As a result, the desired genes often have to be recloned.

On the other hand, if we have a standard set of vector parts that can be combined using a standard set of tools and processes, any two laboratories using these standards will be able to share materials and techniques much more easily. This will mean greater versatility in finding solutions for cloning problems. Most importantly, it will save time and effort.

5.4 The BioBrick Standard

In 2003, Thomas Knight from the Artificial Intelligence Laboratory at the Massachusetts Institute of Technology (MIT) proposed a method of standardizing the structure of DNA fragments in order to standardize their handling and use. He called this the "BioBrick standard for physical composition of biological parts." He proposed flanking DNA fragments with EcoRI and XbaI restriction sites upstream and SpeI and PstI downstream (◘ Fig. 5.1). This constitutes a BioBrick insert. Every insert is carried by a BioBrick vector.

◘ **Fig. 5.1** The figure shows the common features of a basic BioBrick. Note the presence of the two sets of paired restriction enzymes. The EcoRI restriction site is just upstream of the XbaI restriction site and, after a stretch of DNA, the SpeI restriction site is found just upstream of the PstI restriction site. Each BioBrick part comprises a segment of DNA flanked by the EcoRI-XbaI and SpeI-PstI restriction site pairs

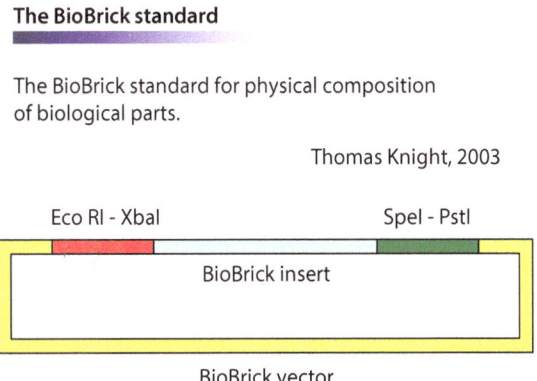

The BioBrick standard

The BioBrick standard for physical composition of biological parts.

Thomas Knight, 2003

Eco RI - XbaI   SpeI - PstI

BioBrick insert

BioBrick vector

5

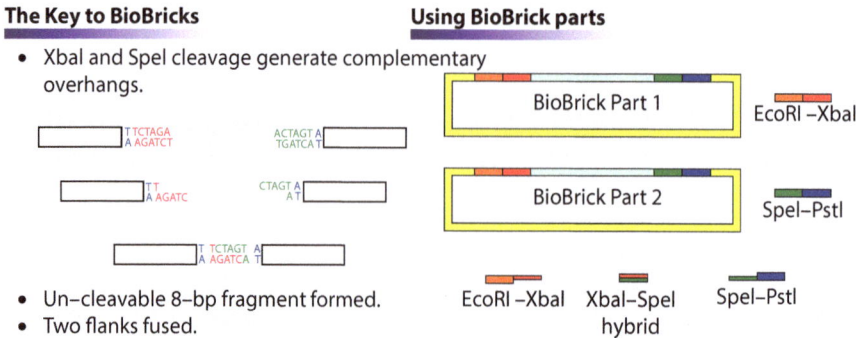

Fig. 5.2 The key to the BioBrick standard – XbaI and SpeI. Restriction of the XbaI (*red*) and SpeI (*green*) recognition sites results in complementary overhangs. This allows a cleaved XbaI site to anneal with a cleaved SpeI site. Once ligated, the resultant hybrid element will no longer be recognised by either XbaI or SpeI

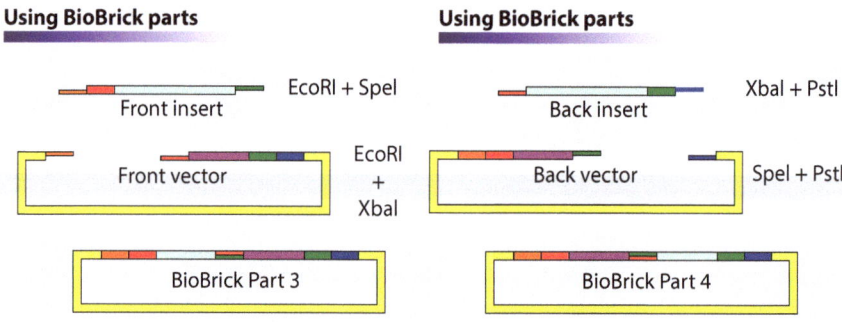

Fig. 5.3 The figure illustrates how BioBricks are used. Basic BioBrick parts can be combined in different ways to produce a more complex BioBrick part, comprising two former BioBrick parts. The choice of whether to use EcoRI or PstI in addition to the special pair of XbaI and SpeI determines whether a BioBrick will be spliced upstream or downstream of the target BioBrick part. Most importantly, the combined parts will be separated by an un-cleavable hybrid XbaI-SpeI element and *still* flanked by the EcoRI-XbaI and SpeI-PstI restriction site pairs. In other words, this new combination of parts is, itself, a BioBrick part

These restriction sites were selected for a special reason—XbaI and SpeI restriction enzymes generate complementary overhangs that can base pair promiscuously. However, once ligated, neither enzyme will be able to cleave the resulting eight-base-pair element (■ Fig. 5.2).

Use of these restriction enzymes, in specific combinations with either EcoRI or PstI, allows one to insert a foreign BioBrick insert upstream or downstream from an existing BioBrick insert (■ Fig. 5.3). Most importantly, the new compound insert will *still* carry EcoRI and XbaI restriction sites upstream and SpeI and PstI downstream. So, it too is a BioBrick part.

Restricting BioBrick Part 1 with EcoRI and SpeI will release a front insert from the donor vector. Restricting BioBrick Part 2 with EcoRI and XbaI will create an acceptor vector. The restriction digestions generate complementary overhangs, which will result in the front vector inserting upstream of BioBrick Part 2. This creates a third BioBrick part.

Conversely, restricting BioBrick Part 1 with XbaI and PstI will release a back insert from the donor vector. Restricting BioBrick Part 2 with SpeI and PstI will create an acceptor vector. The restriction digestions generate complementary overhangs, which will result in the back vector inserting downstream of BioBrick Part 2. This creates a fourth BioBrick part.

In this way, a growing collection of BioBrick parts, each compatible with any other, can be obtained. These could be used and modified just as easily by any other laboratory using the same standard—the BioBrick Assembly Standard 10. Such laboratories could form highly versatile and compatible networks.

5.4.1 Parts, Devices, and Systems

Next, Knight went on to establish the Registry of Standard Biological Parts. This is a special bank where BioBrick material of all kinds is deposited either as sequence information or as physical DNA fragments. All deposits in the registry are organized in different ways. They can be grouped according to function, species, encoded content, and so on. The simplest means is to rank them by complexity into parts, devices, and systems.

A part is basically a fragment of DNA with a defined function, such as a promoter, protein coding sequence, or terminator. One could recombine specific parts to create a higher function. An example would be to recombine BioBrick promoters and coding sequences to create genes. Such a combination of parts is called a BioBrick device. BioBrick devices can, in turn, be recombined to create a collection of related functions, such as in an expression vector where several tasks including host selection, protein expression, and replication are performed. Such a recombination of devices is called a system.

The important thing is that each BioBrick can be combined with any other BioBrick in the registry since they are all produced using the same standards. In this way, even BioBrick devices will be compatible with each other and BioBrick systems will have extremely exchangeable components for increased versatility.

Registered users can freely use the BioBrick parts from the registry so long as the contributor is acknowledged. In turn, registered contributors can add parts to the registry, so long as they adhere to the BioBrick standard and promise not to enforce intellectual rights on registered users. They must also ensure that their parts adhere to BioBrick Assembly Standard 10.

5.4.2 The BioBrick Foundation

In 2005, Drew Endy, who had worked closely with Thomas Knight, founded the BioBrick Foundation. This is an organization that allows the synthetic biology community to share ideas, resources, and activities, including teaching and training. It also organizes the annual International Genetically Engineered Machine (iGEM) competition, where undergraduate students are given actual BioBrick parts and challenged to produce something useful from them.

The goal of this foundation is to create a community wherein synthetic biology is conducted ethically and with sincere aims, in a free, safe, and effective manner. They hope that this will result in work that will benefit mankind and our world in general. It brings together engineers, scientists, lawyers, students, teachers, and laypersons, encouraging them to work toward these ideals.

5.4.3 iGEM

This activity is best represented by the annual iGEM students' course, and later competition, begun at MIT in 2003. Teams are formed and each is given a package of BioBrick parts, which they must use to create novel BioBrick systems. In this way, the students will contribute, through high-quality work, to the development and future of synthetic biology.

■ **Summary**

The BioBrick standard embodies the synthetic biology ideal of applying engineering rigor to biology. Ultimately, the aim is to know as much as possible about the materials used, the tools needed, and the processes employed. This knowledge allows us to control the quality of the work done and will allow standards to be set and met on reliability, ease, versatility, ethical research, and safety.

There are other similar attempts to standardize molecular biology. Just like the BioBrick Assembly Standard 10, each has its advantages and disadvantages. Casini et al. provided a review of these methods in their 2015 article titled "Bricks and Blueprints: Methods and Standards for DNA Assembly."

5.5 Discovery of the CRISPR/Cas Immune System

It was 1987 when Japanese scientist Yoshizumi Ishino stumbled across a very interesting gene locus in *Escherichia coli*. This is a region made up of a series of DNA repeats. These were also observed by Spanish scientist Francisco Mojica, who decided to call the locus SRSR for "short regularly spaced repeats." He later suggested calling it CRISPR, which stands for "clustered regularly interspaced short palindromic repeats" (◘ Fig. 5.4).

◘ **Fig. 5.4**   The figure shows how short palindromic repeats (*red*) are clustered in bacterial genomes to form CRISPR loci. CRISPR is the acronym for "clustered regularly interspaced short palindromic repeats"

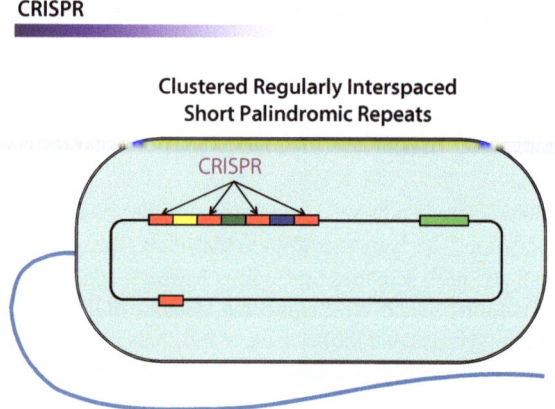

CRISPR

Clustered Regularly Interspaced Short Palindromic Repeats

CRISPR

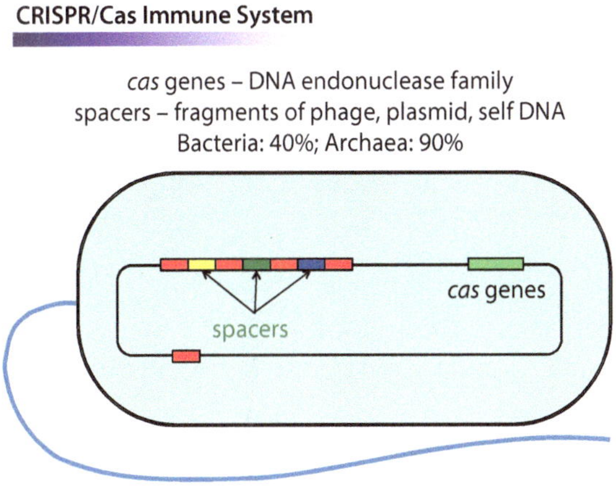

Fig. 5.5 The clustered palindromic repeats in the CRISPR loci are separated by same-sized stretches of distinct DNA elements, called spacers. Although the short palindromic repeats were what first caught the attention of scientists, it is the spacers between them which have come to be of greater interest. Spacer DNA typically comprises fragments of foreign DNA, such as phage genetic material. Associated with the CRISPR locus is another cluster of genes which encode the Cas proteins

Each of these repeats is 24–48 nucleotides long, and they are spaced 20 nucleotides apart. When scientists began looking at sequenced prokaryotic genomes, they found CRISPR loci in about 40% of bacterial samples and in almost 90% of archaeal samples. However, they still did not quite know what its function was.

Then, they started studying the sequences in detail and found that each repeat has a dyad symmetry. These are similar to the DNA recognition sites for type I restriction enzymes. What are even more interesting are the 20 nucleotide sequences *between* these repeats. These are called spacers and are identical to parts of phage DNA or foreign plasmids (**Fig. 5.5**). The idea quickly developed that CRISPR loci might be the heart of some sort of prokaryotic immune system. But how does it work?

By this time, it was also known that near the CRISPR locus is a group of genes encoding enzymes that can unwind and cleave DNA. These are the *cas* genes. The fact that they are associated suggested that the Cas proteins somehow work with the CRISPR locus to defend the prokaryote against reinfection.

What happens, in fact, is that when an invader, such as a phage or another bacterium, injects its DNA into the prokaryote, Cas proteins attack this DNA by binding to it and cleaving it into short fragments (**Fig. 5.6**). These fragments are then carried back to the CRISPR locus and *added to it*. In this way, a new spacer is created between the repeat sequences and a bit of the invader is added to the prokaryote's CRISPR library. A single invader may contribute more than one spacer to the locus. The prokaryote will now be able to recognize the invader in future attacks (**Fig. 5.6**).

It is necessary for the Cas proteins to be able to differentiate foreign DNA, introduced by an invader, from that in its CRISPR library. Otherwise, they would cleave the spacers from the CRISPR locus, leading to considerable genome damage. This is prevented by the need for a 3- to 5-nucleotide signal, found only on the foreign DNA, for the Cas proteins to be active. This signal is the protospacer adjacent motif (PAM). The Cas enzyme will cleave the DNA only if it finds a PAM sequence nearby. This model was confirmed when a group managed to make *Streptococcus thermophilus* immune to phage invasion, using spacer DNA derived from a phage.

Later, when the prokaryote is attacked by the same invader, the CRISPR locus becomes active again. At first, the entire locus is transcribed, so we have a single long messen-

5

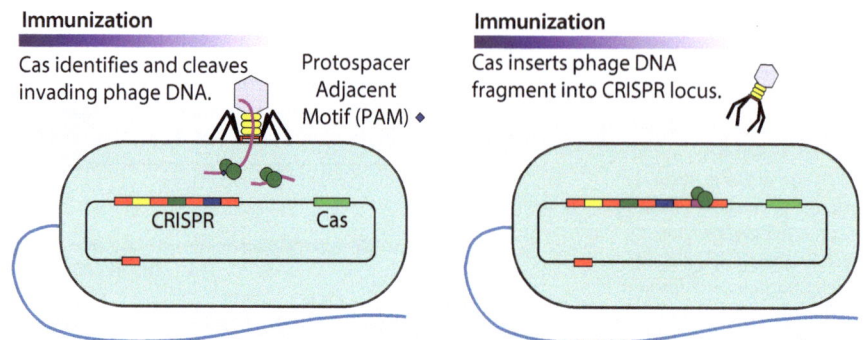

Fig. 5.6   The figure illustrates how the CRISPR-Cas system responds to an invasion of foreign genetic material. Novel phage material is cleaved and incorporated into the CRISPR loci as spacers. These would now serve as identifiers of the invader in the next encounter and primes the bacterium to respond in defence

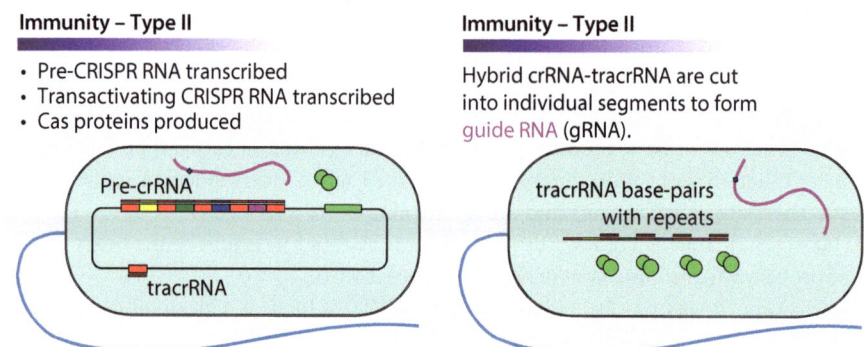

Fig. 5.7   The figure shows a Type II CRISPR-Cas response to a second invasion by the same phage DNA. The entire CRISPR locus is transcribed into pre-crRNA, the tracrRNA is similarly produced and the Cas genes are expressed. Hybridisation of tracrRNA to the pre-crRNA repeats and subsequent cleavage of the double-stranded junctions create guide RNA (gRNA) fragments

ger RNA encoding the repeats as well as the spacers between them (■ Fig. 5.7). This is called pre-CRISPR RNA (pre-crRNA). A second RNA molecule called transactivating crRNA (tracrRNA) is also produced. This is complementary to, and base pairs with, the pre-crRNA repeat sequences. The result is a hybrid single- and double-stranded RNA complex where the single-stranded pre-crRNA segments code for a spacer sequence and the double-stranded segments are the repeat sequences base paired with the tracrRNAs.

This complex carrying multiple spacers will then be cleaved into individual spacer–duplex units (■ Fig. 5.8). Each unit will then act as a guide for other Cas proteins. The latter will unwind and scan the invading DNA for sequences complementary to the guiding spacer sequence. If these are found, and if a PAM signal is at hand, the crRNA–tracrRNA–Cas complex will proceed to cleave the DNA. In this way, the invading DNA will be degraded.

Fig. 5.8 Each gRNA then forms a complex with a Cas protein. Each of these complexes then scans the invading DNA for sequences complementary to their gRNA spacer sequence. When these are encountered, the Cas protein may cleave the invading genetic material. In the Type II CRISPR-Cas system, the Cas protein will only cleave a target if a protospacer adjacent motif (PAM) signal is present. (Modified from CtSkennerton, 2014)

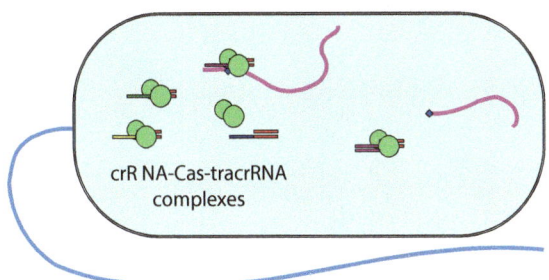

Immunity – Type II

- Cas-gRNA active complexes.
- gRNA recognises spacer sequence.
- Cas cuts if PAM is present.

crR NA-Cas-tracrRNA complexes

Three classes of such CRISPR/Cas immune systems have been defined: types I, II, and III. These three classes differ in how the crRNA–tracrRNA complex guides Cas to deal with the foreign DNA. The type II system would later be developed into the CRISPR/Cas9 technology.

This entire process involves considerable DNA cleavage and ligation. Foreign DNA is first cleaved into fragments, and for these fragments to be inserted into the CRISPR locus, so must the CRISPR locus be. Insertion of the new spacer sequences involves ligation of the DNA material. Cleavage and ligation of DNA are processes at the heart of molecular biology. Can the CRISPR/Cas system be used, then, for genetic engineering? Jennifer Doudna, Emmanuelle Charpentier, and their co-workers, decided to see if this was possible using the type II CRISPR/Cas system (Fig. 5.9).

5.6 Adapting the CRISPR/Cas9 System for Molecular Biology

In the type II CRISPR/Cas system, Cas9 and RNaseIII cut the crRNA–tracrRNA complexes to produce the guide complexes or guide RNAs (gRNAs). These gRNAs then each form an active complex with Cas9. As this complex scans foreign DNA, it will search for sequences complementary to its crRNA. When it encounters the target, it will scan the vicinity of the target for a PAM signal. Having identified such a signal, Cas9 will cleave the target DNA. Up to this point, the important components for this process to work are (1) the target sequence; (2) a PAM signal near the target; (3) crRNA; (4) tracrRNA; (5) RNaseIII; and (6) Cas9.

Suppose you want to cleave the DNA of a cell at a specific site. Suppose, also, that this cell does not carry the CRISPR locus. For any target sequence to be cleave specifically, all one needs to supply are the crRNA, tracrRNA, and Cas9. Of these three, only the crRNA is unique, since it is unique to the target sequence. If one could produce active crRNA–Cas9–tracrRNA complexes carrying different crRNA, one could target multiple sequences simultaneously. What is important is that each sequence must be accompanied by a PAM signal. The most commonly exploited PAM signal is NGG, where N is any nucleotide.

5

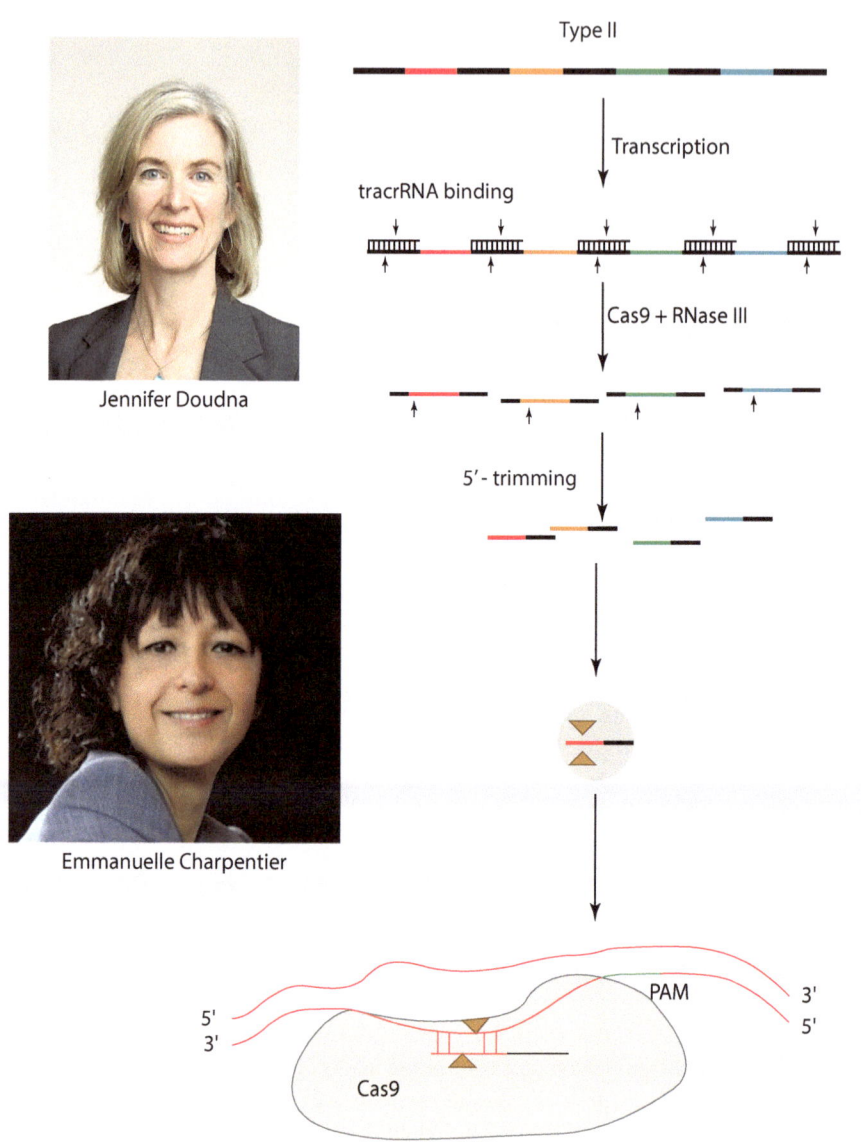

Jennifer Doudna

Emmanuelle Charpentier

◼ **Fig 5 0** The figure illustrates the basic features of the Type II CRISPR-Cas system in greater detail. Of note is the Cas9 protein, which was the first to be specifically developed as a CRISPR/Cas tool for molecular biology. Inset, are two of the scientists who guided this development

Since PAM signals are short and not very specific, it is likely that a target would have one associated with it.

In 2014, Doudna and Charpentier showed that one could introduce the crRNA, tracrRNA, and Cas9 genes into a target cell with a single plasmid encoding all three components. Instead of having the crRNA and tracrRNA separately encoded, one could

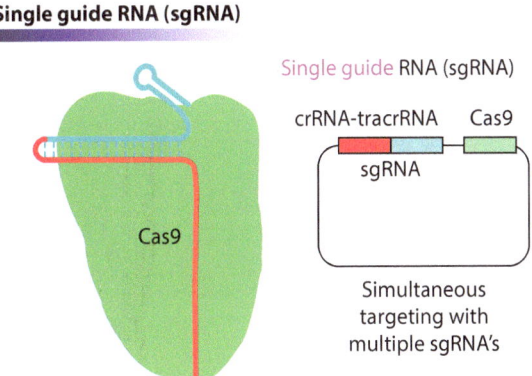

Single guide RNA (sgRNA)

◘ Fig. 5.10 To make the CRISPR-Cas9 system amenable for molecular biology, it had to be streamlined for efficiency. Instead of an entire CRISPR locus, researchers might use only a single spacer for a specific target as the pre-crRNA (now the crRNA). Instead of expressing the tracrRNA separately from the crRNA, the tracrRNA sequence might be spliced downstream of the crRNA. In this way, crRNA and tracrRNA would be inextricably linked, forming a single guide RNA (sgRNA), and ensuring more efficient formation of the active complex. The final component for a working system would be the Cas9 protein. Genes encoding all three components can be carried on a single vector, which can then be transformed into the cell to be modified

also concatenate these two genes to form a single construct—single guide RNA (sgRNA) (◘ Fig. 5.10).

In fact, multiple crRNAs can be included in the plasmid so that the resultant crRNA–tracrRNA–Cas9 complexes can simultaneously target multiple sequences. To demonstrate this, a team in Boston cleaved 62 retroviral sequences simultaneously from the genome of a porcine kidney cell line, without the need to remove and replace this genetic material. This work will be discussed in detail later. Clearly, CRISPR/Cas9 is a powerful new tool for genome editing.

5.7 What Is It Capable Of?

It should be noted, however, that the mainstay of the CRISPR/Cas9 system is not only the ability to cleave specific DNA sequences. It so happens that the Cas9 protein has variants with unique functions apart from cleaving DNA. Some cleave only one strand of their targets, while others cleave both. Some Cas9 variants do not cleave DNA at all. Each variant can be exploited for specific needs. In this way, the CRISPR/Cas9 system can be used for a wide range of genetic manipulations. The following are some examples.

5.7.1 Knock-In and Knock-Out Mutations

When a Cas9 variant capable of cleaving DNA is used, either homology-directed recombination or nonhomologous end joining will be used to repair the damaged DNA (◘ Fig. 5.11). If repair template DNA with flanking homology is provided, this template

The next step – DNA repair

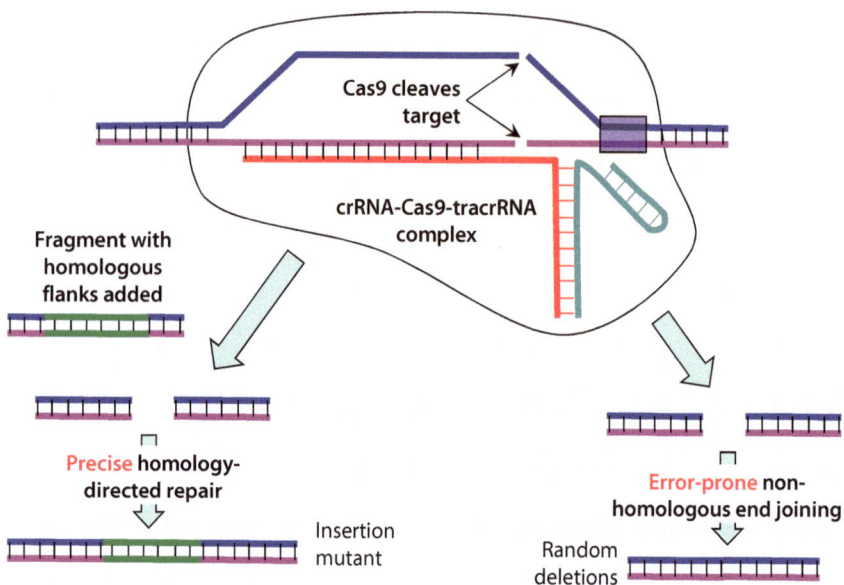

◻ Fig. 5.11 When used to cleave genetic material, the CRISPR-Cas9 system has to be supported by a DNA repair mechanism. Homology-directed repair can be exploited to create controlled insertion mutations in the target genome. In contrast, non-homologous end joining is error-prone and would result in random mutations when the cleaved DNA is repaired

will be incorporated into the cleaved site precisely. This approach can be used to insert a gene into a target genome. This strategy can also be used to permanently disable a target gene by inserting a disrupting nonsense fragment therein.

5.7.2 Gene Regulation

One might even disable a gene only temporarily, if a Cas9 variant which does not cleave target DNA at all is used. An interesting phenomenon observed is that crRNA–Cas9–tracrRNA activity somehow triggers methylation of the target DNA. This will silence the gene but will not destroy it. Once this methylation is removed, the gene will be active again. One can use the same variant of Cas9 but modify it so that it now carries a trans-activating factor. In effect, the complex now has a DNA binding domain and a transactivating domain, just like a transcription factor. Such complexes have been used to upregulate gene expression. Such a system has similarities with the yeast two-hybrid system.

5.7.3 Applying CRISPR/Cas9 to Genome Editing

CRISPR/Cas technology has proven to be a powerful tool in biological research. Several studies have tried to systematically knock out every gene in the human genome, one at a time. This has allowed us to understand the function of each gene as well as its importance, such as whether its function is compensated for by other genes or whether it is resistant to mutation.

In medicine, CRISPR/Cas9 has been used to delete specific genes in order to artificially create known diseases. This has been done in cells as well as in animals, providing us with disease models for study. Use of this technology on cancer cells is allowing us to understand how drug resistance develops in cancer treatment. It was used to disable 62 genes in porcine kidney cells known as porcine endogenous retroviruses (PERVs). PERVs are a major concern when considering the use of pig organs for transplantation. Finally, it is hoped that this technique might someday be a powerful treatment for diseases such as acquired immune deficiency syndrome (AIDS), where in situ gene regulation of the viral genome might be possible.

5.8 What Makes the CRISPR/Cas9 System Unique?

There has been an explosion of publications describing the use of CRISPR/Cas9. However, neither the technology to target DNA nor that to cleave and insert DNA is new. Conventional systems such as Zn-finger nucleases and transcription activator-like effector nucleases (TALENS) can emulate CRSPR/Cas function. Homologous recombination technology is almost as old as the internet. In fact, the Cre–lox recombination system, developed in 1992, is a very similar technology to CRISPR/Cas9, although it has not been as successful. These technologies have been limited largely by the fact that we have to engineer proteins, such as Zn-finger nucleases and TALENS, in order to target a gene. Manipulation of protein structure and function is more difficult than manipulation of nucleic acids. The Cre–lox recombination system requires one to insert *loxP* target sequences on either side of the target in order for the system to recognize it. This means genetically modifying the host *in order to genetically modify it*. In comparison, CRISPR/Cas9 only requires one to produce the guide RNA, and there is no need to modify the target DNA beforehand. To help researchers design gRNAs, developers have created various online software for gRNA design. All of this renders the CRISPR/Cas9 system an attractive alternative for genetic manipulation.

5.9 What Dangers Does It Pose?

The fact that components of CRISPR/Cas9 are so easily created, inexpensive, and easy to use is also what makes the technology a matter for concern. More people will have access to it, and that means more variety in motivations for doing so. So far, we have discussed what biologists and medical researchers hope to achieve with CRISPR/Cas9 technology. What will others, perhaps those outside the scientific mainstream, want to use CRISPR/

Cas9 for? Will it always be for benign purposes? What if CRISPR/Cas9 is used on humans?

CRISPR/Cas technology makes changes that may be permanent in the individual, but only in the cells that have been modified with the crRNA–tracrRNA–Cas9 plasmid. However, if this mutation is made in an important gene locus, it will be preferentially passed on to dividing cells. If this occurs in the sperm or egg cells—the germline—the mutation becomes inheritable. Again, if this mutation is coupled to a critical gene locus, it could be evolutionarily preserved and quickly spread throughout the species, making it a species mutation. This might accelerate human evolution in unpredictable ways. Once again, the concern is human motivation. CRISPR/Cas9 is a very real and—for the first time—very practical tool for creating humans with specific biological advantages. Those who control this technology might become a powerful factor in how society is organized. Humans modified in such ways might also have an unprecedented impact on the environment and existing life on earth. Such hazards will be much harder to manage, since the threat would be human.

Furthermore, powerful as it is, CRISPR/Cas9 is not perfect. In 2015, a group of Chinese scientists caused panic when they reported a successful modification of human embryonic stem cells. What is just as important is the fact that CRISPR/Cas9 made unexpected mutations elsewhere, besides the target site. This happened because of how frequently one finds PAMs in the human genome, and how many genes share sequence homology. At this stage, if this technology is used to modify a germline, we might create offspring with unintended disabilities. Unlike natural disabilities, these would be the result of willful human tampering.

5.10 How Are These Dangers Addressed?

What can be done to manage the hazards of CRISPR/Cas9? To deal with the emerging risks, we need (1) understanding of what CRISPR/Cas9 can or cannot do; (2) knowledge of what has already been tried and what is possible; (3) understanding of what dangers CRISPR/Cas9 could create; and (4) guidelines regulating the application of CRISPR/Cas9, especially to humans. These processes should involve both scientists and nonscientists. Discussions such as this, involving people of diverse backgrounds, have recently become more frequent and more urgent. This is especially so since genome editing was first reported to have been performed on human cells. These discussions would hopefully result in clear guidelines on how CRISPR/Cas9 is used, and for what purpose. Some groups have already firmly advised against the use of CRISPR/Cas9 for specific cases, such as germline engineering.

However, whatever decision is reached regarding the use of CRISPR/Cas9, or indeed any powerful tool, it will only have meaning to those who partake of the agreement. How should we prepare to address the work of those who do not?

┌─ **Take-Home Messages** ───

1. Projects in synthetic biology that call for genetic engineering require methods that are powerful, precise, reliable, and safe.
2. The use of BioBricks allows the process of cloning to be more reliable, simple, and compatible between laboratories.
3. This can be achieved by making functional fragments of DNA, such as promoters and coding sequences, easily interchangeable.
4. Each BioBrick is a fragment of DNA with defined structures at the flanks.
5. Using a specific combination of enzymes, one can ligate a BioBrick upstream or downstream of another BioBrick, as desired.
6. A repository exists for BioBricks, which is accessible to all registered users.
7. The BioBrick Foundation aims to encourage the use of BioBricks among its members.
8. It also aims to develop a community, comprising scientists and nonscientists, which discusses and uses this technology for the betterment of mankind.
9. This work is extended to the annual International Genetically Engineered Machine (iGEM) competition. Here, teams of students and mentors are provided with a collection of BioBrick parts and challenged to produce novel BioBrick devices and systems.
10. The CRISPR/Cas system is thought to constitute the immune system in prokaryotes.
11. This system involves Cas enzymes, which have varying functions ranging from DNA helicase activity to single-stranded or double-stranded DNA cleavage.
12. The Cas enzymes are guided to their target by guide RNAs, which are transcribed from fragments of previously encountered foreign phage or plasmid DNA.
13. The CRISPR/Cas9 system has been exploited for in situ site-specific DNA modification in target cells.
14. These modifications include knock-in and knock-out mutations, gene up- and downregulation, and massive deletions and insertions into target genomes.
15. Unlike conventional methods of in situ genome modification, the CRISPR/Cas9 system does not require the design and production of specialized proteins, nor prior modification of the target genome.
16. Furthermore, CRISPR/Cas9 requires only a plasmid encoding the necessary Cas enzymes as well as the target sequences that would form the guide RNAs.
17. The ease of production and use of components of the CRISPR/Cas9 system makes it readily available, and an attractive alternative, to most researchers, including those who may not be affiliated to official laboratories.
18. As CRISPR/Cas9 has the potential to make significant changes to target genomes, its misuse by any user is of great concern.
19. Modification of genomes in germlines and other important targets might lead to hereditary modifications. This is of concern since it could potentially accelerate evolution of the target species, with unpredictable consequences for human health and the environment.
20. To deal with this risk, there should be public discourse about CRISPR/Cas9, for the purpose of helping other scientists and laypeople to understand the technology and what it is capable of. There should be a conscious effort, involving people of diverse backgrounds, to create guidelines and regulations regarding the use of CRISPR/Cas technology.

Further Reading

Baltimore D, Berg P, Botchan M, Carroll D, Charo RA, Church G, Corn JE, Daley GQ, Doudna JA, Fenner M et al (2015) Biotechnology. A prudent path forward for genomic engineering and germline gene modification. Science 348:36–38

Casini A, Storch M, Baldwin GS, Ellis T (2015) Bricks and blueprints: methods and standards for DNA assembly. Nat Rev Mol Cell Biol 16:568–576

Doudna J (2015) Genome-editing revolution: my whirlwind year with CRISPR. Nature 528:469–471

Doudna JA, Charpentier E (2014) Genome editing. The new frontier of genome engineering with CRISPR-Cas9. Science 346:1258096

Freedman BS, Brooks CR, Lam AQ, Fu H, Morizane R, Agrawal V, Saad AF, Li MK, Hughes MR, Werff RV et al (2015) Modelling kidney disease with CRISPR-mutant kidney organoids derived from human pluripotent epiblast spheroids. Nat Commun 6:8715

Hart T, Chandrashekhar M, Aregger M, Steinhart Z, Brown KR, MacLeod G, Mis M, Zimmermann M, Fradet-Turcotte A, Sun S et al (2015) High-resolution CRISPR screens reveal fitness genes and genotype-specific cancer liabilities. Cell 163:1515–1526

Ho-Shing O, Lau KH, Vernon W, Eckdahl TT, Campbell AM (2012) Assembly of standardized DNA parts using BioBrick ends in *E. coli*. Methods Mol Biol 852:61–76

Jinek M, Chylinski K, Fonfara I, Hauer M, Doudna JA, Charpentier E (2012) A programmable dual-RNA-guided DNA endonuclease in adaptive bacterial immunity. Science 337:816–821

Lanphier E, Urnov F, Haecker SE, Werner M, Smolenski J (2015) Don't edit the human germ line. Nature 519:410–411

Liang P, Xu Y, Zhang X, Ding C, Huang R, Zhang Z, Lv J, Xie X, Chen Y, Li Y et al (2015) CRISPR/Cas9-mediated gene editing in human tripronuclear zygotes. Protein Cell 6:363–372

Merkle FT, Neuhausser WM, Santos D, Valen E, Gagnon JA, Maas K, Sandoe J, Schier AF, Eggan K (2015) Efficient CRISPR-Cas9-mediated generation of knockin human pluripotent stem cells lacking undesired mutations at the targeted locus. Cell Rep 11:875–883

Shalem O, Sanjana NE, Hartenian E, Shi X, Scott DA, Mikkelsen TS, Heckl D, Ebert BL, Root DE, Doench JG et al (2014) Genome-scale CRISPR-Cas9 knockout screening in human cells. Science 343:84–87

Shetty RP, Endy D, Knight TF Jr (2008) Engineering BioBrick vectors from BioBrick parts. J Biol Eng 2:5

Wang H, Yang H, Shivalila CS, Dawlaty MM, Cheng AW, Zhang F, Jaenisch R (2013) One-step generation of mice carrying mutations in multiple genes by CRISPR/Cas-mediated genome engineering. Cell 153:910–918

Yang L, Guell M, Niu D, George H, Lesha E, Grishin D, Aach J, Shrock E, Xu W, Poci J et al (2015) Genome-wide inactivation of porcine endogenous retroviruses (PERVs). Science 350:1101–1104

Zhu W, Lei R, Le Duff Y, Li J, Guo F, Wainberg MA, Liang C (2015) The CRISPR/Cas9 system inactivates latent HIV-1 proviral DNA. Retrovirology 12:22

5

Dealing with the Dangers

© Springer International Publishing AG 2018
E.-K. Ehmoser-Sinner, C.-W. D. Tan, *Lessons on Synthetic Bioarchitectures*, Learning Materials
in Biosciences, https://doi.org/10.1007/978-3-319-73123-0_6

6

What You Will Learn in This Chapter
Particular concerns regarding synthetic biology work will be presented in this chapter. What hazards are involved and what threats they pose will be discussed. We will also look at methods to assess these hazards in order to prepare for any mishaps they might cause. Assessment of risk will be defined in terms of the impact such hazards would have on human health and the environment, as well as the likelihood of such mishaps happening. We will discuss some means by which risk assessments can be made more reliable, as well as some strategies for minimizing the impact of certain hazards.

6.1   The Risks of Synthetic Bioarchitectures

We have seen how different groups in the synthetic biology community have highlighted and addressed safety concerns arising from such work. As with all powerful and fast-moving technologies, synthetic biology—and consequently its toolbox, synthetic bioarchitectures—can have severe and far-reaching impacts on existing life, if not guarded against in time.

> **Potential of Synthetic Biology**
> We are about to learn how to manipulate the most basic elements of living systems
>
> 'In near future, synthetic genomics technology should make it possible to recreate any existing virus for which the complete DNA sequence is known.'
>
> http://www.grid.unep.ch

This is particularly true in the case of synthetic organisms. Globalization has already produced unintended consequences, such as xenobiotics, which are organisms that (even unintentionally) are transported by travelers or as contaminations on containers or ship surfaces (for example, water ballasts). They are often imported because they are considered attractive, such as many flowers—for example, the beautiful *Kosmee* flower. Another example is the large, sweet, colorful *Pomacea* water snail, which became popular in aquaristics but became a considerable threat to Spanish rice fields; consequently, the importation of *Pomacea* has been prohibited by European Union (EU) law since 2013. Invasive organisms, such as this snail, sometimes present a disturbance to the natural balance in a population, and many examples come to mind, from plants to animals, where this phenomenon has been observed. Over time, many such "intruders" have become integrated into the ecosystem; however, in other cases, the original "wild" form has vanished.

There is deep concern that the toolbox of synthetic bioarchitectures could be harnessed for the development of a new generation of life-forms, which could be considered "intrinsic xenobiotics." And, not by chance, the term "synthetic biology" has been strongly linked in history to the term "human enhancement." Humans have already started—sometimes unintentionally—to expose habitats to foreign life-forms that present alien properties to the environment; in some cases such alien properties can lead to an advantage, while some do not. It will constitute an interesting question as to what extent drastic combinations of "nonevolutionary" genetic combinations crossing the natural borders of species will overstretch the Darwinian context.

Synthetic bioarchitectures represents a toolbox for engineering of life—and it would be harder for existing life to adapt to drastically engineered life-forms as they possibly are "outsiders" for the natural inhabitants and additionally they will inherently modify their environment, as any life-form does.

Such contact poses the following hazards: (1) the synthetic organism might out-thrive existing life-forms, leading to depopulation, or even extinction, of the latter; (2) if it survives and spreads, the synthetic organism cannot be removed entirely even if it has escaped from the laboratory environment unintentionally; and (3) the behavior (e.g., feeding habits) of the synthetic organism could be destructive to the environment. Research on ecosystems and modeling of interactions and population dynamics would be of help in order to assess and mitigate potential risks from engineered life-forms, as this is already necessary to understand and protect natural habitats from such xenobiotics.

It is a valid question in synthetic biology as to whether such drastic changes in an organism (such as implanting a foreign genome) can be assessed under the same boundary conditions as those for a "standard" genetically modified organism.

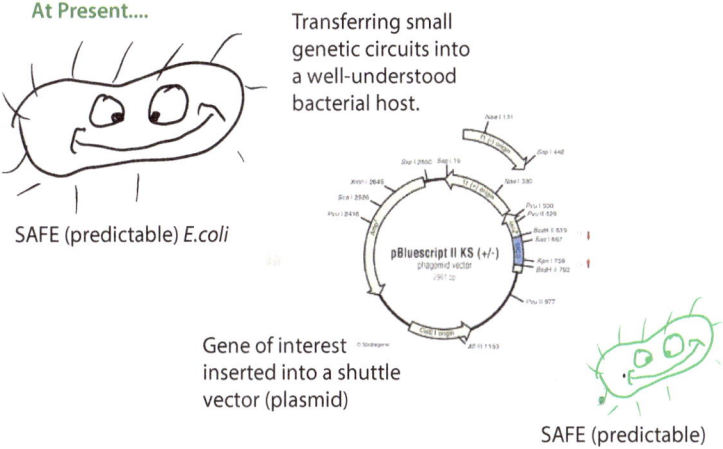

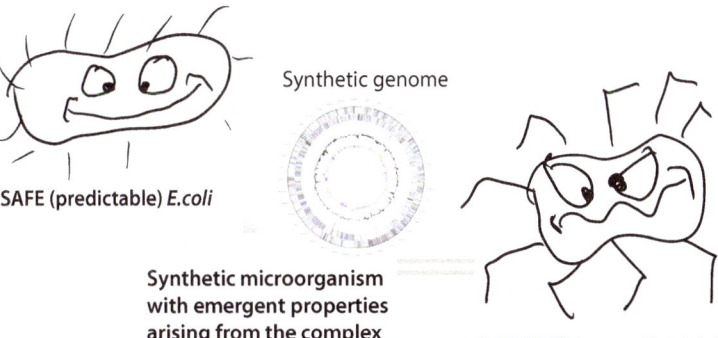

There are three major sources of concern: (1) the ease of access, by improperly informed laypersons, to materials and methods used for synthetic biology; (2) desire on the part of individuals or groups to harness the power of synthetic biology to cause harm; and (3) unforeseen threats to existing life, for example, posed by novel living systems.

HOWEVER...

The "precautionary principle" has to be applied - treating synthetic microorganisms as dangerous until proven harmless.

6

Many of the materials and methods used in synthetic biology are readily available to users even if they are not affiliated with any accredited organization. Various projects already exist that encourage amateur or alternative experimentation with synthetic biology. This raises the question of what motivates such groups and why they operate outside the mainstream.

In summary, even if the intentions of researchers are completely benign, there is still uncertainty as to what the impact of synthetic systems would be on existing life. This is true for both amateurs and bona fide scientists and engineers. A means is needed to guard against unforeseen exposure of existing life to the products of synthetic biology.

Synthetic Biology and the Risk of Deliberate Misuse

The Central Intelligence Agency of the US released a short white paper that concludes,

"Growing understanding of the complex biochemical pathways that underlie life processes has the potential to enable a class of new, more virulent biological agents engineered to attack distinct biochemical pathways and elicit specific effects."

6.2 Risk Assessment

To be effective in preparing for a mishap, a rational approach is required. This requires that (1) a threat be clearly defined, so that it can be recognized; (2) precautions be used so that a mishap is prevented from occurring; and (3) plans be put in place for what to do if such a mishap occurs, so that its consequences can be minimized or removed altogether. Risk is not the same as uncertainty—actually, it is quite the opposite: a risk can be considered only once the procedures involved have been identified and, as such, a risk is involved in something we are familiar with. Uncertainty still can be involved, as our expectations and competence in prediction might be insufficient.

This approach treats every potential mishap as a risk. Risks comprise (1) the dangerous material or process in consideration; (2) the impact of exposure to this material, or occurrence of this process; and (3) how frequent such exposure or occurrence would be. These components allow us to categorize risks and prepare, or act, to mitigate them.

To reduce the subjectivity of this process, a method has been devised to quantify risks, by first assigning values to the impact of a hazard (Consequence) and the frequency, or likelihood, of its occurrence (Frequency). Risk is then equated to the product of Consequence and Frequency:

$$\text{Risk} = \text{Consequence} \times \text{Frequency}$$

The resultant value is then used to judge the severity of the risk, and that, in turn, will suggest what precautions and counteractions to take. In summary, as risk never equals zero, if a facility operates for long enough, it is certain—statistically speaking—that it will experience an accident. In other words, we need to assess and (if possible) reduce risks, and mitigate any consequence of the application of synthetic biology, as we can never surely exclude a potential mishap.

So, we can only assess the bona fide risks according to the "most actual" status of research and technology. To make ourselves familiar with the most actual status and the dissemination of the most recent research results, this is where the "responsibility" of a society—and, finally, the individual—kicks in.

6.2.1 **The Consequence Term**

The matrix below shows how one might grade the Consequence term. Depending on the effects of the mishap, its Consequence can be assigned a value.

For example, a hazard whose effects are to (1) cause a single, disabling injury to the worker; (2) result in hospitalization or serious injury to others; (3) cause irreversible damage to the environment; and (4) cost $1–10 million in damages annually will be assigned a value of 3, which represents a Consequence described as "severe". Such matrices are already standard in organizational portfolios in order to assess the risk of certain processes in monetary values.

Consequence Matrix

	Employee Safety	Public Safety	Environmental Impact	Economic value
low	Report	No impact	Limited impact (correctable)	10–100k€
moderate	Hospital	Small impact (smell, etc)	Report to Agencies	100–1M€
severe	Disabling injury	Hospitalization /reports in public	Irreversible Damage	1–10M€
very severe	Letal/multiple severe injuries	Letal/multiple severe injuries Massive negative publicity	Catastrophic consequences	Larger than 10M€

Adapted from: Ian Sutton, Process Risk and Reliability Management, Elsevier, 2015, 2nd Edition

6.2.2 **Frequency of Occurrence**

A similar matrix can be drafted for Frequency. In this way, Frequency can also be assigned a value.

The Consequence and Frequency matrices can then be combined to give the Risk matrix. In this way, the degree of risk posed by a threat can be determined, and appropriate precautions and actions can be arranged. For instance, a risk that is judged to be low might warrant the placement of equipment or materials, such as first aid kits, which the endangered might use to assist themselves. A risk that is judged to be high might warrant the use of highly specialized protective equipment or perhaps the establishment of a specialized team of helpers who are trained to handle the specific hazard.

Use of the Risk matrix suggests that no hazard can be completely free of consequences or can never occur. This means that any hazard can never be risk free, although one can try to reduce the impact and likelihood of a hazard through planning and preparation.

6.2.3 **Caveats**

Not all hazards may be assessed by the same determinants and thresholds. Furthermore, not enough information is available sometimes for Consequence or Frequency matrices to be used objectively.

> **Facts are never truly objective**
> A truth ceases to be a truth as soon as two people perceive it.
> *Oscar Wilde (1854 – 1900)*

In such cases, the values chosen might be influenced by assumptions, estimations, beliefs, and other biases. This often exacerbates the subjectivity of risk assessments or, as Ian Sutton stated in his famous book: "We only learn from blood" (see "Further Reading").

Involvement of Subjectivism into Risk Assessment

Sutton Ian

- The subjective component of risk becomes even more pronounced when the perceptions of non-specialists, particularly members of the public, are considered.
- Hence the successful risk management involves understanding the opinions, emotions, hopes and fears of many people, including managers, workers and members of the public.

6.3 Model Ecosystems

As another means of reducing this element of subjectivity, models might be used to study hazards in miniature. This allows the assessors to observe the impact of a hazard on a particular set of conditions, on a small scale. An example would be to use isolated model ecosystems to study the impact of synthetic life on existing flora and fauna.

Furthermore, one might ask if the risks posed by synthetic organisms are any different from those posed by genetically modified or even nonindigenous organisms. From the perspective of general biosafety, all three groups pose the same risks. Any uncharacterized organism, including synthetic ones, should be handled as a health hazard until proven otherwise.

6.4 Handling Biohazards

Following the precept favoring prevention over a cure, general biosafety stresses containment of hazardous living material. This means to limit or prevent exposure of existing life to the hazard unless proper precautions are in place. This approach would be just as effective when used with synthetic organisms.

Living material is typically handled in biosafety facilities. Experimental cell and tissue materials should be considered capable of causing human disease or environmental damage when released. As such, they should be handled in isolation according to the appropriate biosafety level. A synthetic organism, by virtue of being novel and uncharacterized, should be considered highly suspicious and therefore warrant more stringent biosafety precautions.

In case the synthetic organism manages to escape a biosafety facility, it should be ensured that it would not survive, nor have a selection advantage over native flora and fauna. This can be done by designing a biological response, such as suicide, following specific conditions such as overgrowth or exposure to an introduced environmental signal.

The Risk of Accidental Release

"Despite the fact that no accidental release of a genetically- engineered microorganism (GEM) from a laboratory has been reported, it is possible that such releases have occurred but that the effects were so unremarkable that they remained undetected."

THE NEW ATLANTIS, JONATHAN B. TUCKER AND RAYMOND A. ZILINSKAS
www.TheNewAtlantis.com

6.5 Orthogonality

Orthogonality is one of the aims in synthetic biology—for example, when it comes to pathway engineering. The adjective "ὀρθός" (*orthós*) comes from Old Greek and means "correct" or "right"; together with the suffix "-gonal" it describes the right angle and is used in the language of mathematics. The right angle is the least interference of a vector with

6

another—describing the attempt to disturb a system as little as possible. Transferring this term to molecular biology, it means to "reprogram" a gene without any disturbance of the context. Everyone who has ever tried to knock out a single gene knows how "nonorthogonal" this attempt becomes in most cases! The complexity and still unknown regulation factors and intermingled pathways often lead to unexpected consequences of the resulting organism. In plant biology, the multiple genomes add to this problem. So, we can understand the aim "orthogonality" as an aim for the synthetic biologist, which is addressed in systems biology, when the network of pathways is the subject of research and identification of individual "valves" in a respected pathway is the prerequisite for successful pathway engineering, as the mutation in this special place will most likely lead to a controlled change and, thus, can be termed "orthogonal" as only the desired pathway is affected by the change. Still, many side effects happen, as interconnected pathways are regulated by products and educt concentrations are changed. This is an inherent bottleneck in modifying a living organism; however, in so-called cell-free approaches, one can investigate such changes in a cellular lysate and once the regulatory interactions are resolved in such systems, the chances for investigation of isolated gene products become higher.

In another context, such orthogonal changes are valuable tools for mitigating the risk of unintended release of synthetically engineered organisms. One strategy to prevent the survival of such released organisms is so-called biological firewalls—for example, the use of noncanonical amino acids in a modified organism. Several strategies are underway, as such synthetic organisms might be designed to subsist only on synthetic nutrients, such as nonnatural DNA and amino acids. This strategy works only if the modified pathway results in an "orthogonal" mutation, leading to a 100% dependent species, which has no work-around strategy in place to bypass the critical pathway, so that essential supplements must be supplied for them to remain viable. Such measures will ensure that these organisms will not survive outside the laboratory. Such special metabolic requirements would also ensure that synthetic organisms would not be able to influence existing life biologically. For example, a synthetic organism designed to use xenonucleic acids (instead of nucleic acids) for genetic material would not be able to affect natural life genetically. Such a strategy was also mentioned in ▶ Chap. 2, where autopoiesis was discussed. If the synthetic organism is designed to be viable only under non-natural conditions, such as the use of non-natural nutrients, exposure to the natural environment should lead to its death, preserving the natural inhabitants.

6.6 Constant Monitoring

Transparency and careful scrutiny of research work, such as monitoring of the purchase of hazardous material, will allow watchers to determine if suspicious research activity is present. This is particularly useful if such material is being accessed by amateurs or rogue scientists. Mainstream research work, on the other hand, can be easily curbed through control of resources, such as funding and facilities. This is one of the most critical issues in synthetic biology and its toolbox—synthetic bioarchitectures. Here we need to rely, as scientists and members of our respective societies, on transparency of and access to research results, and governance by the laws of the respective country. As a consequence of terrorist attacks, the acquisition and use of critical genetic sequences is monitored by the companies and databases involved. This is one of the attempts to restrict and control research, and the same strategy is in place for stem cell use. Still, the deliberate misuse of genetic material/information is one of the critical threats of today's societies.

■ Conclusion

Ultimately, it needs to be those who are responsible for the products of synthetic biology who are the most vigilant. However, their concerns and recommendations must reach others, particularly those who might be affected, as well as those involved with regulation of research. This can only be effective when there is constant, open, informed, and rational dialogue across all social strata.

Take-Home Messages

1. Synthetic biology results in materials and living organisms that may turn out to be hazards.
2. The impact that a hazard has on human and environmental health can be mitigated if one is prepared.
3. Preparation requires that hazards be identified before they cause a mishap.
4. Once identified, the risk they pose can be evaluated by considering the severity of the effects of any mishap on human and environmental health, as well as how frequent or likely such mishaps might be.
5. Risk assessments are inherently subjective, since different people might perceive severities or likelihoods differently. Various means are needed to reduce this subjectivity.
6. The impact of a hazard on the environment can be studied in model environments. These models mimic important characteristics of the environment without exposing the environment to the hazard.
7. Synthetic organisms should be handled in the same way as any uncharacterized naturally occurring organisms. In the absence of contrary evidence, both kinds of organisms should be considered dangerous to human health and the environment (the precautionary principle).
8. Constant monitoring and evaluation of research activities, particularly in the area of synthetic biology, would allow potentially hazardous work to be dealt with in time.
9. Public discussion of controversial work in synthetic biology will help to maintain vigilance. This would be necessary in order to prepare for any hazards that may arise.

Further Reading

Brander R (1995) The Titanic disaster: an enduring example of money management vs. risk management. Essay on risk management. Calgary Unix Users Group, Canada. Available online at www.cuug.ab.ca/Bbranderr/risk_essay/titanic. html (Accessed 20 Oct 2013)

CDC-NIH (2009) Biosafety in microbiological and biomedical laboratories, 5th edn. Atlanta, USA: U.S. Department of Health and Human Services

Doudna J (2015) Genome-editing revolution: my whirlwind year with CRISPR. Nature 528:469–461

Schmidt M (2008) Diffusion of synthetic biology: a challenge to biosafety. Syst Synth Biol 2:1–6

Schmidt M (2010) Xenobiology: a new form of life as the ultimate biosafety tool. Bioessays 32:322–331

Sutton I (1997) Process safety management. USA: Southwestern Books

Sutton I (2008) Use root cause analysis to understand and improve process safety culture. Process Safety Progress 27(4):274–279

Sutton I (2014) Process risk and reliability management, 2nd edn. USA: Elsevier

WHO (2004) Laboratory biosafety manual, 3rd edn. Geneva, Switzerland: World Health Organisation